Handbook
of
Power Generation
Transformers
and
Generators

John E. Traister

Prentice-Hall, Inc., Englewood Cliffs, N.J. 07632

Library of Congress Cataloging in Publication Data

Traister, John E.   (date)
  Handbook of power generation.

  Includes index.
    1. Electric power production.   2. Electric generators.
  3. Electric transformers.   I. Title.
  TK1001.T73   1983   621.312   82–20480
  ISBN 0-13-380816-5

Editorial/production supervision: Gretchen Chenenko and Barbara Palumbo
Interior design: Gretchen Chenenko
Cover design: 20/20 Services Inc.
Manufacturing buyer: Anthony Caruso

229797

Printed in the United States of America
10 9 8 7 6 5 4 3 2 1

ISBN   0-13-380816-5

Prentice-Hall International, Inc., *London*
Prentice-Hall of Australia Pty. Limited, *Sydney*
Editora Prentice-Hall do Brasil, Ltda., *Rio de Janeiro*
Prentice-Hall Canada Inc., *Toronto*
Prentice-Hall of India Private Limited, *New Delhi*
Prentice-Hall of Japan, Inc., *Tokyo*
Prentice-Hall of Southeast Asia Pte. Ltd., *Singapore*
Whitehall Books Limited, *Wellington, New Zealand*

Contents

Preface

The subject of electrical power generation and distribution has been covered by many books from the standpoints of both a formal presentation of complex theory and the presentation of standardized information on common problems. Few, however, are written in such a way as to allow a technician or engineer to immediately design and select a suitable system for a practical application without first wading through complex theory that typifies most other books on the subject.

The intent of *Handbook of Power Generation: Transformers and Generators* is to dwell only briefly on theories and then immediately progress into practical, on-the-job applications that are used for almost every possible situation. It is one of the first books to show the reader what equipment is available and how to use this equipment for specific applications.

Employees of power companies and industrial plants will find this book indispensable and will want to keep it close at hand for frequent reference. Consulting engineers will want a copy in their library, as will all electrical technicians and engineers. Students taking electrical engineering courses should find this book a very helpful supplement to their textbooks dealing mainly with theory. In fact, everyone involved in the electrical industry in any capacity should find this book a welcome addition to their reference library.

As with any book as comprehensive as this, it is not possible for one person to create such a work. Rather, it is a compilation of the efforts of many. The response from manufacturers, engineers, and power company personnel was overwhelming, and without their help, this book could not have been written. We have included most of those responsible in the acknowledgments section.

John E. Traister

Acknowledgments

The author consulted several sources when preparing the material for inclusion in this book. Besides using the books and other publications listed, the author consulted and drew material from technical magazines and journals, trade-association standards, engineering and scientific papers, industrial and engineering catalogs, and a variety of similar publications. Acknowledgments are as follows: National Fire Protection Association; Square D Company; Onan Corporation; Gould, Inc.; Hevi-Duty Electric; McGraw Edison Company; Rapid Electric Company, Inc.; Acme Electric Corporation; R. E. Uptegraff Manufacturing Company; Sylvania; General Electric; Westinghouse Corporation; Cummins Engine Company, Inc.; Honeywell, Inc.; A. B. Chance Company; Frequency Technology, Inc.; and Emerson Electric Company.

Credit for additional material appears at various locations throughout the book.

1

Introduction
to Power Generation

The earliest recorded practical use of commercial electricity was the installation of an arc lamp in the Dungeness lighthouse in England in 1862. This early type of light, while not entirely steady or free from smoke, was able to produce great amounts of very bright light by drawing electric arcs or flames between two carbon electrodes. However, it was not until 17 years later, in 1879, that the first successful carbon-filament incandescent lamp was invented by Thomas A. Edison. He also developed the first efficient electric generator to supply electrical current for his lamps in the same year. The experiments were continued, and in 1882, Edison developed the first central-station electric generating plant in New York City.

Prior to Edison's generating plant, naturally there was no inclusion of electrical systems for residential, commercial, or industrial applications. However, this new generating plant brought on a public demand for the use of electric lighting and power for all types of applications and was actually the beginning of power generation.

Since Edison's time, the electrical industry has grown to become one of the largest in the world, and as the systems became more comprehensive and complex, the need for more qualified electrical engineers and technicians became necessary, requiring much study into the development and design of power generation and distribution systems.

GENERATORS

The heart of most electrical systems is the electric generator which produces electric power by generating a voltage due to electromagnetic induction. In its simplest form, a generator operates as shown in Fig. 1-1. The two single-turn coils (A, B, C, and D) shown are arranged to revolve in the field of permanent magnets (drawings a and b). Note that the ends of the coils are connected to metal slip rings which are fastened to the revolving shaft and, consequently, turn with it. This arrangement provides a connection from the moving coils to the resistance load by means of metal or carbon brushes rubbing on the slip rings.

If the coil (A, B, C, D) revolved clockwise, wire A, B will be moving upward through the flux of the magnetic field, and the induced voltage will be in the direction indicated by the arrow down on the wire. In the same drawing, wire or side C, D is moving downward, and its induced voltage will be in the reverse direction but will join with and add to that of wire A, B, as they are connected in series in the loop. Note that the current flows to the nearest collector ring and out along the lower wire to the lamp, returning on the upper wire to the farthest collector ring and the coil.

Drawing b in Fig. 1-1 shows the same coil after it has turned one-half revolution farther, and now wire A, B is moving downward instead of up as before. Therefore, its pressure and current are reversed. Wire C, D is now in the position where A, B was previously, and its current is also reversed, flowing out to the farthest collector ring, and over the top wire to the circuit load, and returning on the lower wire.

From the preceding brief explanation of a simple generator, we see that as the conductors revolve through the flux of the magnetic field—passing first a north pole and then a south pole—their current is rapidly reversed. Therefore, the current produced by such a generator is called *alternating current.*

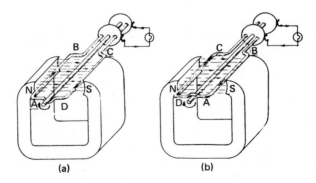

(a)   (b)

FIGURE 1-1 Simple electric generator of one single wire loop in the flux of a strong permanent magnet. Drawing 2 shows that the coil has revolved one-half turn farther.

If it is desired to obtain direct current (nonreversing current), some type of commutator or rotary switch must be used to reverse the coil leads to the brushes as the coil moves around. In other words, all generators produce alternating current in their windings and must be converted as described previously if direct current is desired.

A single-loop generator with simple commutators, for producing direct current, is shown in Fig. 1-2. Here again we have a revolving loop. In drawing a in Fig. 1-2, wire A, B is moving up, and its current is flowing away from the commutator—and that of C, D toward the commutator. Note that the coil ends are connected to two bars or segments of a simple commutator, each wire to its own separate bar. With the coil in this position, the current flows out at the right-hand brush, through the load in the circuit, and reenters the coil at the left brush.

In drawing b of Fig. 1-2, the coil has moved one-half turn to the right, and wire A, B is now moving down, and its current is reversed. However, the commutator bar to which it is connected has also moved around with the wire, so the current still flows in the same direction in the external circuit through the load in the circuit.

When it is necessary to operate direct-current equipment from alternating-current systems (away from the generator), a rectifier is normally used. With such a device, dc equipment can operate from a conventional ac supply. The fundamental principles of rectifier operation are explained in Chapter 5.

TRANSFORMERS

The electric power produced by alternators in a generating station is transmitted to locations where it is utilized and distributed to users. Many different types of transformers play an important role in the distribution of electricity. Power transformers are located at generating

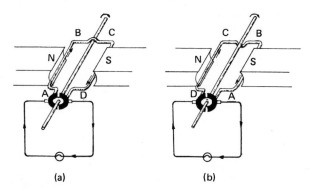

(a)   (b)

FIGURE 1-2 Single-loop generators with simple commutators for producing direct current.

stations to step up the voltage for more economical transmission. Substations with additional power transformers and distribution transformers are installed along the transmission line. Finally, distribution transformers are used to step down the voltage to a level suitable for utilization.

Two coils or windings on a single magnetic core form a transformer. Such an arrangement will allow transforming a large alternating current at low voltage into a small alternating current at high voltage or vice versa. Transformers, therefore, enable changing generators which produce moderately large alternating currents at moderately high voltages to very high voltage and proportionately small current in the transmission lines, permitting the use of smaller cable and providing less power loss.

The elementary principle of a transformer is shown in Fig. 1-3. Two windings are shown on a rectangular core made of iron. The source of alternating current and voltage is connected to the primary winding of the transformer. The secondary winding is connected to the circuit in which there is to be a higher voltage and smaller current, although it could be a larger current and smaller voltage. If there are more turns on the secondary than on the primary winding, the secondary voltage will be higher than that in the primary and by the same proportion as the number of turns. The secondary current, in turn, will be proportionately smaller than the primary current. With fewer turns on the secondary than on the primary, the secondary voltage will be proportionately lower than that in the primary, and the secondary current will be that much larger. Since alternating current continually increases and decreases in value, every change in the primary winding of the transformer produces a similar change of flux in the core. Every change of flux in the core, and every corresponding movement of magnetic field around the core, produces a similarly changing voltage in the secondary winding, causing an alternating current to flow in the circuit which is connected to the secondary.

Theoretical study of conditions in a transformer includes the use of phasor diagrams which represent graphically voltages and currents in transformer windings. Calculations of impedance can be simplified by using equivalent circuits. Reactance voltage drop is governed by the leakage flux, and the voltage regulation depends on the power factor of

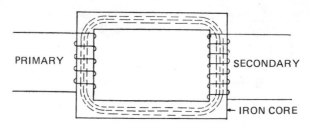

FIGURE 1-3 Core and windings of a simple transformer.

the load. To determine transformer efficiency at various loads, it is necessary to first calculate the core loss, hysteresis loss, eddy-current loss, and load loss. The methods to be applied are explained in later chapters in this book (see also the Appendix).

Other areas covered in later chapters include transformer connections, parallel operation of transformers, voltage regulation, and testing of transformers, all of which are necessary for proper transmission systems design and maintenance.

Distribution transformers are used in transmission and distribution systems to provide the desired quantities of voltage. Most are rated up to 500 kVA, have steel cores of the core or shell form, and use cylindrical coils for voltages up to about 25,000 V; disk coils are used for higher voltages. Distribution transformers are protected by primary fuses, lightning arresters, surge diverters, and proper grounding. They are available as residential pole-mounted or pad-mounted transformers, rural-line transformers, and unit substations which contain all auxiliary equipment and may be stationary or portable.

Power transformers are designed for higher voltages and kVA ratings than distribution transformers. Their cores are constructed either in shell or in core form, and the high-voltage windings generally consist of disk coils. The shielding of coils helps to improve voltage distribution under voltage surges, while temperature indicators and thermal relays give a warning or disconnect the transformer when the temperature rises too high within the transformer. Almost all power transformers are oil filled and can be self-cooled, forced-air cooled, water cooled, and forced-oil cooled, depending on specific conditions. Transformer oil needs careful attention because of expansion, breathing action, deterioration due to moisture, and fire and explosion hazards. Inert-gas systems and activated alumina filters protect the oil. Because of many auxiliary devices, power transformers need careful maintenance. A trouble chart in Chapter 13 shows the most common operational troubles and their remedies.

There are also several miscellaneous transformer types that will be fully covered in this book. For example, those with only one winding are known as air-core or iron-core reactors. Constant-current transformers are built to provide constant current regardless of load variations. You will also find transformers designed for specific applications such as control transformers, test transformers, and the like.

ELECTRIC TRANSMISSION

When designing an electric transmission system, the following points should be considered:

1. The conductors must be so proportioned that the energy transmitted through them will not cause an undue rise of temperature.

2. The conductors must have such mechanical properties as to enable them to be successfully erected and so durable as to require a minimum of annual maintenance.

3. The conductors may be so chosen that the initial cost of line construction will be at a minimum, that the station construction will be a minimum, that the cost of the plant will be reduced and the cost of operation and maintenance will be at a minimum, that the total cost of the installation will be at a minimum, that good service will be at a maximum, and that the initial cost of the generating plant will be at a minimum, consistent with maximum income.

The success of any electric transmission system requires that conditions 1 and 2 should be met. However, it is not possible to fulfill all the conditions of 3—but designers should strive for a proper balance of these conditions.

Outside electric transmission systems are installed either overhead or underground. The former includes such work as wood pole overhead line construction, steel tower overhead line construction, substation and switchyard construction, and overhead trolley systems. Underground facilities are either buried directly in the earth or the conductors are pulled through an underground raceway system. Auxiliary equipment such as capacitors, lightning arresters, overcurrent protection, relays, and the like are included in such systems.

The size of the conductor to obtain economical transmission depends on the strength of current; that is, the lower the current rating, the more economically it may be transmitted. Therefore, a given quantity of energy can be transmitted much cheaper at high voltage than at a lower voltage. Because of the readiness with which ac systems may be transformed (up or down) by means of transformers, practically all long-distance electric transmission is by alternating current.

To transmit a given quantity of energy with a given percent drop of potential and a given loss, the weight of the conductor varies inversely as the square of the voltage and directly as the square of the distance.

POWER PLANT OPERATION

In general, there are three kinds of electric power plants in normal use. Alternate systems such as wind and solar are still undergoing refinement and are not yet suitable for commercial applications.

FOSSIL FUEL Fossils fuels (including geothermal) supply about 80% of the electricity used in North America. Burning coal, gas, or oil heats water to steam, which spins a turbine to turn generators, converting heat energy into electrical energy.

HYDROELECTRIC The pressure of falling water—such as caused by a major dam—spins a turbine, which in turn operates generators, converting mechanical energy into electrical energy.

NUCLEAR This type of power plant produces electricity in much the same way as a fossil fuel plant except the furnace is called a reactor, and the fuel is uranium.

In a typical fossil fuel power plant, coal is carried by conveyor to a hopper-pulverizer where the coal is ground to the fineness of talcum powder. This fine coal is then diverted into a furnace where it is ignited to heat pure water which is circulated in pipes to make steam. The steam generated turns a turbine, which turns the generator rotor to produce electricity. Spent steam is routed to a condenser where it is converted back to boiler water for another cycle.

When the generator turns to produce electricity, the current usually varies between 22,000 and 26,000 V. For long-distance transmission, however, the initial current is passed through a transformer where the voltage is stepped up, sometimes as high as 765,000 V, depending on the distance it must be transmitted. At points of usage, substations are used to step the voltage down to around 12,000 V for local distribution to residential, commercial, and industrial facilities. More transformers are used at points of usage, which further step the voltage down, the most common ones being 480/277 V and 240/120 V.

Advances in technology will bring new methods of producing electrical energy. For example, in the not-too-distant future, *fusion* may be used to produce electricity; in fusion, hydrogen nuclei merge and produce great quantities of energy with little if any pollution. Solar and wind energy will certainly find commercial use in the near future. Magnetohydrodynamics (the process where fiery fuel is shot through a magnet to produce electricity) is also under development.

New methods of electric transmission are also under development. For example, higher voltages may be used in transmission lines for greater economy, reliability, and flexibility. You may also see supercooled underground transmission cables emerge shortly.

SUMMARY

The material contained in this chapter is meant to give an overview of power generation and transmission. In nearly all cases, only a basic, general description is given; details will come later, as well as practical applications. With this overview, the reader also has a preview of what is to follow in later chapters: Each phase of power generation (with emphasis on generators and transformers) will be covered in full to provide a complete, single source of information in this field for the worker, engineer, and student.

2

Principles
and Characteristics
of dc Generators

Direct-current power is widely used in many industrial applications. Two important uses are found in electrochemical and industrial processes that require the precise control obtained with dc motors. The industrial processes that use direct current include pulp or paper mills, steel rolling, newspaper printing, and many others which employ tandem drives for close speed control. The characteristics of dc motors make them especially suitable for loads that are difficult to start, where the speed must be varied over a wide range, and where the load must be started and stopped often, such as traction work, milling machines, mine work, lathes, pumps, steel mill work, elevators, and those named previously.

While much dc power is obtained from alternating-current sources and then rectified, in this chapter we shall deal only with direct-current generators. Rectifiers and converters will be covered in Chapter 5.

DC GENERATOR PRINCIPLES

Direct-current electric generators are machines that change mechanical energy into electrical energy and are usually rated in watts or kilowatts. Any dc machine may be used as a motor or generator. Therefore, information that pertains to one applies to the other as well.

The frame of a basic dc generator is made of iron to complete the magnetic circuit for the field poles. Frames are made in three types:

open, semienclosed, and closed. The open frame has the end plates or bells open so the air can freely circulate through the machine. The semienclosed frame has very small holes in the end bells so that air can enter but will prevent any foreign material from entering the machine. The enclosed type, as the name implies, has the end bells tightly closed so that the machine is airtight. Some generators are even watertight, making them suitable for operation under water. Other uses of the enclosed-type generators are in industrial applications where small dust particles are constantly in the air that would damage the generator if allowed to enter.

The field poles are made of iron, either in solid form or built of thin strips called laminations. The iron field poles support the field windings and complete the magnetic circuit between the frame and armature core.

Bearings are used to support the armature shaft and armature and are usually made in three different types: sleeve, roller, and ball bearings.

Oil rings are usually another part of the basic dc generator machine. Most are used in conjunction with sleeve-type bearings and act to carry the oil from the oil well to the shaft. The oil ring must turn when the machine is operating; otherwise the bearing will burn out.

The rocker arm supports the brush holders. On most machines, this arm is adjustable to enable the shifting of the brushes for better operation. When the brushes are rigidly fastened to the end bell, the entire end bell assembly is shifted to obtain best operation.

The brush holders support the brushes and hold them in the proper position on the commutator. The brushes should be spaced equidistantly on the commutator when more than two sets of brushes are used. When only two sets are used, they will be spaced the same distance as a pair of adjacent field poles. The brush tension spring applies enough pressure on the brush to make a good electrical connection between the commutator and brush.

Brushes used on dc machines are made of copper, graphite, carbon, or a mixture of these materials. The purpose of the brushes is to complete the electrical connection between the line circuit and the armature winding.

Commutators are constructed by placing copper bars or segments in a cylindrical form around the shaft. The copper bars are insulated from each other and also from the shaft by mica or similar insulation. An insulating compound is used instead of mica on small commutators. The commutator bars are soldered to, and complete the connection between, the armature coils.

The armature core is made of thin sheets of laminated iron pressed tightly together. The laminated construction is used to prevent induced currents (eddy currents) from circulating in the iron core when the machine is in operation.The iron armature core is also a part of the magnetic circuit for the field and has a number of slots around its entire surface in which the armature coils are wound.

The armature winding is a series of coils wound in the armature slots, and the ends of the coils connect to the commutator bars. The number of turns and the size of wire are determined by the size, speed, and operating voltage of the machine. The purpose of the armature winding is to set up magnetic poles on the surface of the armature core.

The field windings are made in three different basic types:

1. Shunt
2. Series
3. Compound

Shunt fields have many turns of small wire, and series fields have a few turns of heavy wire. The compound field is a combination of the two windings. The name of the field winding depends on the connection with respect to the armature winding. The field winding, of course, produces magnetic poles that react with the armature poles to produce rotation.

Most dc generators use some of the current generated in the armature to supply excitation current to the fields, and this type is appropriately called a *self-excited* generator. However, some are constructed so that the field coils are connected to an outside source of electricity and are known as *separately excited* generators. In this design, when the armature rotates in the magnetic field, current is supplied to the load.

SERIES GENERATOR

The circuit of a series generator is shown in Fig. 2-1. Note that the armature, fields, and load are all connected in series. Therefore, should the load be disconnected from the generator terminals, the circuit

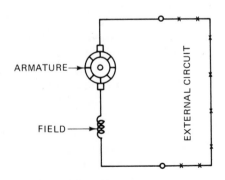

FIGURE 2-1 Diagram of basic dc series generator.

through the generator will be open, preventing current from flowing through the field coils, and consequently no voltage will be generated.

One of the unique characteristics of a series-wound dc generator is the increase of output voltage as the load increases. In other words, if a small load such as one dc lamp is connected to the terminals of a series-wound generator, a small current will flow through the generator, creating a small magnetic flux and a low voltage. As more lamps are added to the circuit, creating a heavier load, a greater current will flow, and consequently more lines of force will be produced, causing a higher voltage to be generated. The voltage at no load is zero, and it increases to a maximum at full load. Of course, there is a limit to the amount of load that may be connected to each generator; it may not exceed the rating of the generator.

SHUNT GENERATOR

The field coils of the shunt generator are connected in parallel, as opposed to series, across the armature terminals as shown in Fig. 2-2. This connection provides for constant field strength regardless of the load. There is one problem however. As the load is increased, the terminal voltage will decrease because of an increased voltage drop within the armature. In other words, the voltage at no load is maximum and decreases slightly as the load is increased.

COMPOUND GENERATOR

The series-parallel or compound generator encompasses several types of dc generators—the short-shunt cumulative type probably being the most common. Figure 2-3 shows that the shunt field is connected across the armature, and the current flow in the shunt field is in the same direction as in the series field.

The short-shunt compound generator usually supplies constant voltage regardless of the load, but its voltage regulation can be varied by changing the number of turns in the series field winding or by using a resistor across the series field to vary the current through it. A device that varies the current through the series field is called a *diverter*.

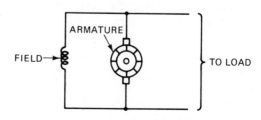

FIGURE 2-2 Field coils of a shunt generator are connected in parallel.

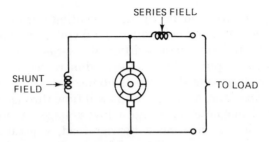

FIGURE 2-3 Shunt field in a compound generator is connected across the armature, and the current flow is in the same direction as in the series field.

Three types of compound generators may be obtained by changing
the number of turns in the series field. These three basic types are called

1. Overcompounded generator
2. Flat-compounded generator
3. Undercompounded generator

The following characteristics are the results of changes in the number of
turns in the series field:

1. If the turns in the series field are increased over the number necessary to
give the same voltage output at all loads, the generator will be overcompounded,
causing the voltage to increase as the load increases. This characteristic is highly
desirable when the generator is located some distance from the load as the rise in
generated voltage compensates for the voltage drop in the line.

2. A slight decrease in the number of turns in the series field produces a
flat-compounded generator. In this type of generator, the voltage produced at full
load will be the same as that produced at no load. Such a generator is useful when it
is located near the load it is "pulling."

3. If the number of turns in the series field is further reduced, it becomes
an undercompounded generator producing normal voltage at no load. As the load
is increased, the voltage drops considerably, until at full load the voltage may be as
high as 20% below normal. Some welding machines utilize this winding as it is
useful where a short circuit may occur.

INTERPOLES

Interpoles are normally used on the generators mentioned. These poles
are connected in series with the armature and arranged so that the
polarity of the interpoles is the same as the main pole ahead of it in the
direction of rotation. When interpoles are used, usually six wires (leads)
are brought out of the generator.

Voltage regulation of dc generators has traditionally been accomplished by the insertion of a field rheostat in the shunt field circuit.
Such an arrangement enables the current to be varied, which in turn

varies the lines of force. With full current in the field, maximum voltage will be obtained. As resistance is added, the generated voltage will become less.

PARALLEL CONNECTION
OF COMPOUND GENERATORS

When a load on a generator exceeds the capacity of the generator, it becomes necessary to either decrease the load or to connect another generator in parallel with the other one, thereby dividing the load between the machines.

To obtain good parallel operation, it is necessary that each of the generators paralleled will run, with any constant load, at a constant speed. An extreme example of a class of generators which can be operated in parallel only with difficulty is one driven by a single-cylinder, single-acting, low-speed engine with a flywheel so small that the speed fluctuates periodically.

When the speed increases, the generator voltage increases and in turn increases the load on this generator and decreases that on the other paralleled generators. When the speed drops, the generator's voltage and load drop, and the load on the other generators increases. Such a generator then will cause the load to surge back and forth between the machines.

Generators driven by turbines or by electric motors do not have this characteristic and so can be paralleled with greater ease. Generators driven by reciprocating engines or steam, gas, oil, or gasoline engines should be viewed with suspicion, although if the speed fluctuations are small and on high frequency, such generators will parallel satisfactorily. In all further discussion it will be assumed that each generator considered is driven at a speed that is constant when a given load is applied. Of course, decrease of speed of the generator due to increase of load also affects parallel operation by causing the voltage to decrease as the load increases. This will be allowed for in all cases by using the volt-ampere curve for a generator obtained by driving the generator by the proper motive power, rather than by driving it at a constant speed.

The second requisite for a generator that is to be operated in parallel is exceedingly important and is the most frequently overlooked. If this second necessary condition to parallel operation were always fulfilled, about 90% of the paralleling trouble would be eliminated. The condition can be stated in many ways. One way of expressing it is, "Each of the generators to be paralleled must tend to shirk its load"; that is, it must tend to transfer its load to the other machine. It may be stated in more detail as follows: "Each of the generators to be paralleled must be so designed and adjusted that as the current between the points of paralleling and through its armature increases, its voltages as read at the points of paralleling must decrease materially." The points of paralleling will be defined as the point at which the current flowing through the armature of the particular generator divides from the rest of the load current. In most but not all cases, the points of paralleling are the points where the armature leads connect onto the equalizer, line, or bus. If the drooping volt-ampere characteristic required is not obtained, a fair division of load may be obtained for a short time, because an unstable condition is set up, and this load division will change even though the total load does not; one generator would soon probably carry all the load and might even drive the other generator as a motor.

The third requisite to satisfactory parallel operation is the relative shape of the saturation curves of the generators being paralleled. To understand the possible division of load between units, the no-load saturation curve of each should be plotted on a common graph by plotting armature terminal voltage against percent field current. The field current necessary to obtain the rated voltage to be used is 100% for each generator. When the curves are plotted and they appear to have the same general shape, it will be possible to make the other required adjustments and obtain satisfactory division of load. If the units do not have similarly shaped saturation curves, then steps should be taken to alter the main poles of the generators having less saturation. These recommendations can be obtained from the manufacturers.

If the preceding conditions are fulfilled, the generators will each carry without shifting such loads as they may be adjusted to carry. Ordinarily, one adjustment only is made on a shunt generator, that of the shunt field. This is usually a permanent adjustment. Strengthening the series field of a generator has little effect on the division of load at light loads but makes that generator take a larger share of the load at the heavy loads.

GENERATORS CLOSE TOGETHER

Parallel operation systems may be roughly divided into two classes. The most ordinary one is where the generators paralleled are all so closely grouped, as in a power plant, that the resistance of leads, buses, and equalizers is negligible. The second class is where the generators are so far separated that the resistance of the line connecting them appreciably affects the division of load, and the resistance of any size of equalizer that is practicable to use becomes so great as to make the equalizer useless and so causes it to be omitted. The latter case is very common in mine railway systems and on installations where the load division is affected by the change in load at the point of paralleling.

It is probably well to call attention again to what the point of paralleling can do to upset a satisfactory paralleling setup. As referred to above, it is the point of division load in a paralleling circuit at which the load current of a unit divides from the rest of the load on the system. From the several applications for generators operating in parallel it is quite common to have an individual load carried by the unit most likely to be kept on the line. When other generators are paralleled for added load, we find the distribution center for the total load to be divided such that load conditions are not equally divided by distance between armatures and are subject to load shifts from the one point to the other. When the same conditions for division of load are insisted upon, then adjustments are necessary as load shifts.

Considering only systems of the first class, it is evident that at all times each generator must impress the same voltage at the points of paralleling as does every other generator of the system. Also it is evident that if volt-ampere curves are drawn for each generator at the proper excitation, including both shunt and series ampere turns, then on any curve the current read at the voltage found at the points of paralleling must be the current that particular generator is delivering, and the sum of such currents for all the generators must equal the load current. It follows that the separate generator currents, the load current, and the voltage all readjust themselves until these conditions prevail, for no other conditions are stable.

APPLICATIONS OF SHUNT GENERATORS

In Fig. 2-4 are shown typical curves of two shunt-wound generators. Curve A is for a 50-kW, 250-V generator with its shunt field adjusted to give rated voltage and current. Curve B is a similar curve for a

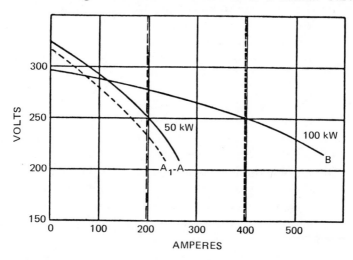

FIGURE 2-4 Typical curves of two shunt-wound generators. (Courtesy Westinghouse.)

100-kW, 250-V generator. It is evident that when two machines so adjusted are paralleled through leads of negligible resistance and a total load of 600 A is applied, each machine will carry its rated load at 250 V. However, with a load of 300 A, generator A would carry 125 A at 281 V, and generator B would carry 175 A at 281 V. If the field rheostat of generator A is changed until the new curve produced, A_1, is such that when 300 A of total load is applied each generator will carry half of its rated current, then when 600 A of total load is applied, one generator will carry more and one less than its rated current, generator A carrying 175 A at 244 V and generator B, 425 A at 244 V.

If the two generators (Fig. 2-4) adjusted to give curves A and B are paralleled and carry a total load of 600 A, each generator carrying its rated current, and then through a momentary change of speed the small generator carries 250 A, its voltage will drop to 225 V while the voltage of the other machine rises momentarily, and the generator with the higher voltage will take an increased load, while the generator with the lower voltage will drop part of its load. This redistribution of load raises the voltage of the generator with the lower voltage, and decreases that of the other generator, until finally each carries steadily its rated current.

If the generators had characteristics as shown in Fig. 2-5 instead of Fig. 2-4 and the load becomes momentarily 250 A on generator A and 350 A on generator B, the difference of voltage between the two generators would become 7 V instead of 33 as in the preceding case. It is obvious that the generators producing the greatest corrective voltage (i.e., those of Fig. 2-5) would be the first to redistribute their load properly and so would be the most stable. Furthermore, in the case of the generators represented by Fig. 2-5, a slight change in a generator has an extremely great effect on the division of load. A slight change in the resistance of the field due to its heating up might change the one generator's curve from A to A_1. Under this condition the load would shift until generator B carried almost all the load.

If two shunt-wound generators are paralleled by means of leads of negligible resistance, we find the following:

1. If each generator has a curve that is decidedly drooping, the generators may be satisfactorily paralleled.

2. The more drooping the curves, the more stable will be the division of load between the generators.

3. The more drooping the curves, the less will the division of the load be affected by slight changes in a generator's curve.

4. The generators may be adjusted by means of field rheostats to give any division of load at any total load desired, but this division will be different at all other loads.

Two compound generators, properly equalized, will operate well in parallel. It should be noted that, considering the effect of its own

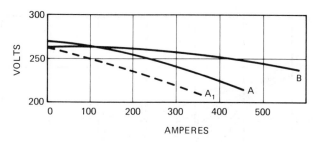

FIGURE 2-5 Less drooping curves of shunt generators.

individual load only, each generator has a drooping volt-ampere curve. The generators then operate in parallel very much as do shunt generators, and all the conclusions drawn concerning shunt generators apply except the fourth. By making adjustments of the field rheostats and of the series resistances, a compound generator can be made to divide the load properly at two different loads instead of one. Usually the adjustments are made for full load and for no load. This is the result obtained in the parallel adjusting procedure given.

The importance of one thing concerning shunt-wound generators, and equally true of equalized compound generators, must not be forgotten. The volt-ampere curve taken at the points of paralleling must be not only drooping but must be decidedly drooping. A few generators, particularly large compensated ones, may be found that do not naturally have such a decided droop. Before paralleling, they must be changed to have such a droop. In general, if two compound generators will operate satisfactorily in parallel as shunt generators, with their series fields cut out, then they should also operate in parallel satisfactorily as equalized compound generators when they are correctly adjusted.

GENERATORS AT A DISTANCE

Two or more generators situated a considerable distance apart are often connected to the same line and operate in parallel supplying power to that line. This happens frequently in mining service, where one generator will be located at the mouth of the mine and the other far in the mine. The generators are so far separated that equalizers cannot be effectively and economically used. The operation of such generators is worthy of study.

PARALLEL GENERATORS

Assume that the two shunt-wound generators whose curves are given in Fig. 2-6 are connected to the opposite ends of a line whose total resistance, including the resistance of the wire in both sides of the line, is

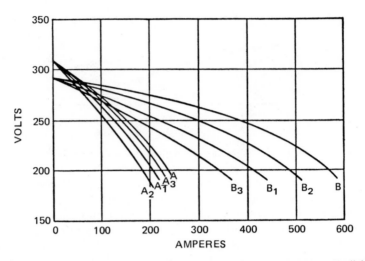

FIGURE 2-6 Regulation curves of two shunt-wound generators paralleled through lines having appreciable resistance. (Courtesy Westinghouse.)

.2 Ω, distributed uniformly with respect to distance. Assume that with a total load of 600 A applied at a point midway between the generators the field rheostat of generator A is so adjusted that each generator carries its rated amperes. Curves A and A_1 of Fig. 2-6 show the regulation curves of generator A when so adjusted, A being taken at the generator and A_1 at the point of paralleling, which in this case is the center of the load or the midpoint of the lines. Curves B and B_1 are similar curves for generator B. Curves A and B are of theoretical interest only. The curves at the point of paralleling, that is, curves A_1 and B_1, are the ones that determine how the generators will parallel. It may be observed that these curves are even more drooping than those of Fig. 2-4, and so the generators may be expected to parallel even better than when located close together. This would be true if the load were constant in position. If the load should move to a point three times as far from generator A as from generator B, the new curves at the point of paralleling would be A_2 and B_2 (Fig. 2-6). With a 600-A load applied at this point, generator A will carry 165 A, and generator B will carry 435 A, the voltage at the points of paralleling becoming 221 V. If the load should move to a point three times as far from generator B as from generator A, the new curves at the points of paralleling would be A_3 and B_3 (Fig. 2-6). With a 600-A load applied at this point, generator A will carry 230 A, and generator B will carry 370 A. It is seen that although the resistance of the line makes the division of load more stable, and causes it to change less with a change of the amount of load, the resistance also causes the division of load to change as the position of load changes. As the load approaches either generator, that generator takes a larger and larger portion of the load.

CHARACTERISTICS OF COMPOUND GENERATORS

Compound generators are more frequently used for this type of paralleling than are shunt generators. Figure 2-7 shows two compound generators so paralleled.

An independent load is also shown as being placed on generator B. Such an independent load may or may not exist. If the generators A and B are the same as those whose curves are shown in Fig. 2-8, if the generators are not equalized, if the resistance of the line is negligible, and if there is no independent load, then the curves of the generators taken at the points of paralleling would be the same as the curves of Fig. 2-8. Since there is no line drop, the voltage at the point of paralleling, or center of load, is evidently the same as the terminal voltage of each of the generators. The curves are rising, not drooping, and as has been shown, the division of load would not be stable.

Assume next that each of these generators is connected to the ends of a line of .2-Ω total resistance and that the load is midway between the generators. The new curves obtained at the points of paralleling are curves A_1 and B_1. If a total load of 600 A is applied at the midpoint, it can be seen that the voltage at that point will be 217 V, and generator A will be overloaded, carrying 288 A, while generator B will be underloaded, carrying 312 A. It is further seen that the curve of generator B taken at the points of paralleling is decidedly drooping. The division of load, therefore, probably would be stable, although not as stable as it would be if curve A_1 were more drooping. However, the division of load is not that desired. It would be desirable, therefore, to shunt current from the series field of generator A until curve A_1', read at the points of paralleling, is obtained. This curve is such that when a load of 600 A is applied at the middle of the line, each generator will carry its rated current. This curve, like curve B_1, is decidedly drooping. When so adjusted, and the load concentrated at the middle of the line, the generators should operate excellently in parallel.

When the generators are adjusted for the load at the midpoint of the line and the load changes its position, the generator which it

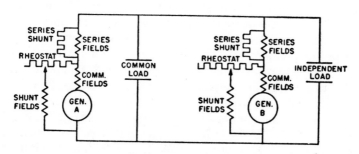

FIGURE 2-7 Connections for compound generators in parallel.

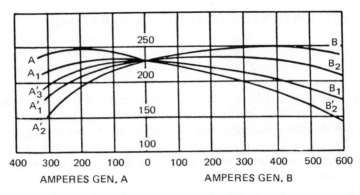

FIGURE 2-8 Regulation curves of two compound generators paralleled through lines having appreciable resistance. (Courtesy Westinghouse.)

approaches will take more than its share of the load. If the generators are adjusted to give curves A_1 and B_1 when the load is midway between the generators, and the load should then move until it is three times as far from generator A as from generator B, the new curves at the points of paralleling would be curves A_2' and B_2' of Fig. 2-8. With 600 A applied at this point, generator A will carry 90 A, and generator B will carry 510 A. If next the load should move to a point three times as far from generator B as from generator A, the new curves at the points of paralleling would be curves A_3' and B_2' of Fig. 2-8. With a load of 600 A applied at this point, generator A will carry 285 A, and generator B will carry 315 A.

When the independent load shown in Fig. 2-7 is applied, the problem becomes more complicated. Generator B is expected to carry all the independent load and part of the common load. When both loads are on and the two loads equal the combined rating of the two generators, then each generator is expected to carry its rated load. If in the case just considered the common load were only 500 instead of 600 A and the independent load were 100 A, then the shunt on the series field of generator A should be adjusted so that generator A carried 200 A of the common load and generator B carried 300 A of it.

Another effect of the independent load should be noted. If with one value of independent load applied the two generators are adjusted to give the desired division of the common load, then when the magnitude of the independent load changes, the division of the common load will change. If generator B is undercompounded at the generator, an increase of the independent load will cause the voltage from generator B to decrease not only at the generator but at all points on the lines. The result will be that generator A will take a greater and generator B a lesser portion of the common load. If generator B is overcompounded at its terminals, then the reverse is true. If it is flat compounded, then a change of the independent load has no effect on the division of the common load.

In paralleling two compound generators for operation with loads as shown in Fig. 2-7 and under the conditions described, the steps taken in paralleling should be as follows:

1. Determine the average point of common load. This need only be approximate.

2. Determine the no-load and full-load voltages desired at that average point of common load (points of paralleling). The greater the decrease of voltage with increase of load, the more constant will be the division of load.

3. Determine the division of common load desired and the magnitude of the independent load to be used in adjusting.

4. Connect generator A to the line, and disconnect generator B.

5. Adjust the shunt field rheostat of generator A to obtain the desired no-load voltage. No common or independent load is then applied.

6. Apply the rated current load of generator A at the average point of common load, and adjust the shunt on the series field of generator A until the voltage read at the point of load is the desired full-load voltage read at the point.

7. Remove the load. Disconnect generator A from the line. Connect generator B to the line.

8. Adjust the shunt field rheostat of generator B to obtain the desired no-load voltage.

9. Apply the independent load determined in step 3. At the average point of common load, apply the portion of the full common load that generator B is to carry. Adjust the shunt on the series field of generator B to such a value as to obtain the desired full-load voltage at the average point of common load.

10. Put generator A back on the line, increase the common load to its full-load value, and check the division of load.

VOLTAGE REGULATION

It is obvious that a shunt-wound generator and a compound-wound generator cannot be made to parallel satisfactorily for a given length of time. There are cases where large units are involved and the cost of change and time involved would be out of question. The change can be offset at a lesser cost by applying voltage regulators equipped with 1R drop compensation that will give regulator response for division of load.

When applied to dc generators, the voltage regulators are, of course, used to secure voltage characteristics and paralleling results that cannot be readily obtained from the generators themselves. When regulators are applied, the generators no longer present their natural characteristics but follow the dc system straight-line characteristic that is imposed on them by the regulating system.

The problems in dc systems are varied and complex and often require adjusting and readjusting and sometimes give rise to this question, "How can it be done otherwise?" In many cases the voltage regulator provides an easy adjustment in making two or more dissimilar generators share their load.

Figure 2-9 shows a characteristic curve of a generator with the characteristic curve of the regulator system superimposed upon it.

The flat curve A represents an ideal voltage regulation of a system without droop up to a full load of 400 A.

Curve B is typical of a shunt generator having an approximate droop of 8%.

Curve C would be typical of a regulated system when controlled by a voltage regulator and having an approximate droop of 4%.

When a fixed point of paralleling is established and a regulator using a differential type of cross-current compensation is applied, the system regulation is corrected to that shown in curve D.

Curves BC and BD are shunt regulation curves of the generator and demonstrate how the regulator shifts the natural regulation of the generator by changing the shunt field excitation.

It is understood from the former discussion that generators located far apart and having a shifting load center are hard cases to adjust. These same generators with regulators follow the same rules as when no regulators are applied. When the load center is fixed, better results are secured. An example of paralleling a group of compound-wound generators and a shunt-wound generator with a fixed load center is shown in Fig. 2-10. An actual case of this kind gave good load division.

Generators A, B, and C were 200-kW generators operating close together. Generator D was a 750-kW shunt-wound unit at considerable distance from the group of compound-wound units.

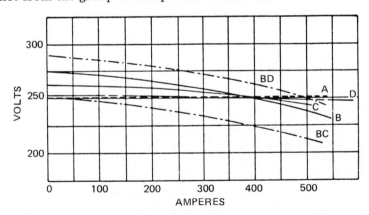

FIGURE 2-9 Characteristic curve of generator and regulator system. (Courtesy Westinghouse.)

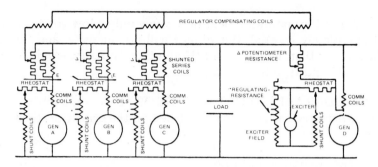

FIGURE 2-10 Compound-wound and shunt-wound generators in parallel with individual voltage regulators. (Courtesy Westinghouse.)

It is concluded that voltage regulators can be used to secure results that could not otherwise be obtained. They make it fairly easy to adjust for various machine differences and system characteristics, but they cannot cure all the troubles that can occur on dc systems of this kind.

TROUBLESHOOTING DC GENERATORS

Generators, like any mechanical device, will develop problems that must be diagnosed and corrected. A complete study of this procedure would require an entire book, if not volumes, but a small sampling is presented here of some of the most common problems.

FAILURE TO GENERATE   If a generator fails to generate, test for loss of residual magnetism. If the field poles lose residual magnetism, it is impossible for the armature to cut lines of force, and therefore no current can be generated. To correct this situation, connect the shunt field to a source of direct current for a few seconds.

The problem could also be due to too much resistance in the field circuit, preventing sufficient current from flowing in the field coils to increase the flux. Such high resistance may be due to the field rheostat, an open circuit in the field, loose connections, poor brush contact, or broken brush pigtails.

Look for an incorrect field connection. The residual magnetism in a generator produces lines of force from a north pole to a south pole. If the current in the field coils is in the wrong direction, lines of force will be produced opposite to the residual lines, and a cancellation of flux will result which will prevent the generator from building up voltage. If this is the case, merely reverse the shunt field connections or reverse the direction of rotation of the generator.

Wrong direction of rotation will have the same effects as reversed field polarity because it causes the current in the shunt field to flow in the wrong direction. To correct this problem, reverse the direction of rotation or interchange the shunt field loads.

A shorted armature or field may allow only a low voltage to build up. If completely shorted, the voltage will not increase, and the armature will more than likely smoke. If all other faults are eliminated, test the armature and field for shorts.

CONSIDERABLE VOLTAGE DROP If considerable voltage drop is experienced when a load is placed on the generator, the trouble may be

1. Differential connection
2. Overload
3. Shorted armature

FAILURE OF VOLTAGE BUILDUP When the rated voltage of a generator fails to build up to maximum, the trouble may be

1. Wrong brush position
2. Shorted armature or field coils
3. Resistance in the field circuit
4. Speed of generator too low

SPARKING BRUSHES Sparking at the commutator is a common occurrence, and one of the chief causes is poor brush contact. This problem may be caused by

1. Worn brushes
2. Clogged brush holder
3. Insufficient spring pressure
4. Loose pigtail connection
5. Brushes shaped improperly
6. Rough commutator
7. Dirty commutator

3

Principles
and Characteristics
of ac Generators

In the early days of the electrical construction industry, direct current was the most popular form of power. However, the distance over which direct current could be transmitted was limited. In contrast, alternating current permits efficient transmission of huge blocks of power between distant cities and generating plants. Therefore, alternating current soon became the principle type of electric power.

ALTERNATING CURRENT

To review how alternating current is produced, let's look at Fig. 3-1. Note that a wire loop—rectangular in form—is arranged to rotate between two powerful magnets and that each end of the wire loop is connected to *slip rings* A and B. Assume that the rectangular loop of wire rotates on its own axis 3600 times/min or 60 rev/sec in a clockwise direction as viewed from the slip rings. Remember that invisible magnetic lines of force (flux) pass horizontally between the two magnets on either side of the wire loop. If the flux lines passed from the left magnet (pole) to the right pole, then at the instant the wire is in the position shown in Fig. 3-1 the current in the loop would flow from front to back. Also at this precise instant of time, the wire is cutting the maximum number of magnetic lines of force, thus generating the highest voltage and current. Simultaneously the wire partially concealed by the right pole is moving through the same flux field but in the opposite

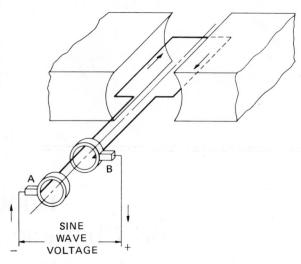

FIGURE 3-1 Simple generator with revolving wire loop arranged in a position so that current flows from front to back. (Courtesy Square D.)

direction. This causes the generated current to flow from the back to the front. Thus the currents are in the same direction in the loop and assist one another. At this instant in time, slip ring A is negative and B is positive.

Figure 3-2 shows the wire loop turned one-quarter of a revolution in which the top and bottom conductors are passing between lines of flux. When the conductors are in this position, no voltage or current is being generated.

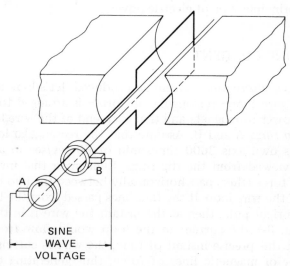

FIGURE 3-2 Here the wire loop is turned one-quarter of a revolution. (Courtesy Square D.)

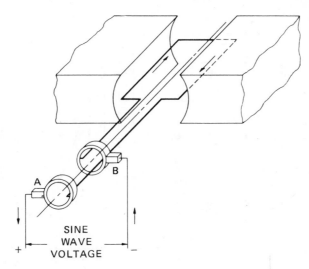

A B

SINE
WAVE
VOLTAGE
+   −

FIGURE 3-3 One-quarter of a revolution later, the current still flows in the same direction as before, but the slip rings have alternated from positive to negative and vice versa. (Courtesy Square D.)

One-quarter of a revolution later the conductors will appear as shown in Fig. 3-3. This position is similar to that in Fig. 3-1, but there is a difference. The current flows in the same direction as before, but slip ring B is now connected to the left side, changing it from positive to negative. Slip ring A has been changed from negative to positive.

Another quarter turn will place the conductors in the position shown in Fig. 3-2, and again no voltage or current is generated. At any point between no voltage (Fig. 3-2) and the highest voltage (Figs. 3-1 and 3-3), the magnitude of the generated current (or voltage) goes from zero to high and from high to zero 120 times every second. The net result of this phenomenon creates a wave, as depicted in Fig. 3-4. In this diagram, the circled numbers 1, 2, and 3 represent the voltage generated in Figs. 3-1, 3-2, and 3-3 respectively. Obviously, there are theoretically an infinite number of positions in any revolution, but for our purposes we shall select positions only every 30° around the full circle of rotation. When these points are plotted, they create a smooth curve that is known as a *sine wave*.

The maximum value or height of the voltage wave is dependent on the constant speed of generator rotation and the strength of the magnetic field. For example, if the electric distribution system was designed for 2400 V between the high-voltage wires out on the transmission lines, the plant operator could increase the magnetic field current enough to boost the voltage on the plant bus to 2450 V. This would assure the operator that out at the end of a long distribution line the last customer's transformer would receive a voltage adequate to provide the required voltage within the limits set by the Public Service Commission.

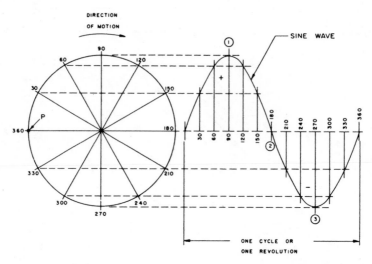

FIGURE 3-4 Diagram of a sine wave depicting one complete cycle.

The majority of electric transmission systems in the United States utilize a 60-cycle (60-Hz) wave which is produced when the generator rotates 120 times/sec to produce 60 identical sine waves, as shown in Fig. 3-4. Hence, a cycle is one complete wave, which when it begins to repeat itself is the start of a new cycle.

When alternating current was first put into operation, many systems utilized 25-Hz generators, but this led to many problems, especially in electric lamps. Each time the current passed to zero, the lamp would try to go out, and this constant flicker was annoying. At 60 Hz, no flicker is discernible.

OPERATING CHARACTERISTICS OF ALTERNATORS

To understand the operation of an actual generator of alternating current (most often referred to as *alternator*), let's look at a simple alternator driven by a gasoline engine—the type normally used for standby power in homes and other types of buildings. The engine has a speed governor, which is a mechanical device that is sensitive to engine speed and, in turn, controls the engine throttle. An increased load on the engine tends to slow it down, so when the governor senses this, the throttle is opened to admit more gasoline to the carburetor to offset this tendency to slow down. This keeps the alternator revolving at the same speed. When the load is removed from the alternator, the engine will pick up speed, but the governor then closes the throttle to bring the engine speed back to normal. Thus the unit operates at rated frequency while supplying a varying load.

The magnitude of the voltage induced in a coil is proportional to the number of turns in the coil, the amount of magnetic flux linking the coil, and the rate at which the flux linkages change with respect to time. For a particular alternator, the number of turns in the coils, and the rate of change of the flux linkages with the coils, are fixed if the machine is operating at its rated speed. But the amount of flux linking the coils can be changed by varying the strength of the magnetic field. This is done by changing the amount of current flowing through the field coils.

Although the voltage induced in the armature winding is directly proportional to the amount of flux, the flux is not directly proportional to the exciting current over the full operational range of the alternator. This is due to the fact that the iron portions of the magnetic circuit present greater reluctance to flux as the concentration of flux, or flux density, becomes higher. At high flux densities the iron becomes saturated by the magnetic field. Because of the saturation of the iron, greater increases in the magnetizing force or exciting current are required to achieve a given increase in flux.

VOLTAGE CONTROL

Since the voltage of an alternator depends on the speed of the machine and on the value of the field current, voltage is controlled by varying the field current by means of a field rheostat or by varying the field current of the exciter. The speed cannot be varied, since this would change the frequency of the system.

Voltage regulators sense the alternator voltage (much the same as the governor senses the speed of a gasoline engine), compare it with a signal representing the desired voltage, and adjust the alternator field current as required to obtain the desired voltage. In years past, this regulation has been performed by electromechanical devices, but they are being rapidly replaced by electronic devices on all modern installations. The static circuitry, in combination with static exciters, has the advantage of exceedingly high response to voltage changes, making it possible to hold voltages more nearly constant through fluctuations in generator load. They also have the advantage of less maintenance and less sensitivity to environment.

LOAD AND POWER-FACTOR REGULATION

Alternator voltage, as well as speed, tends to drop when a load is connected. To offset this tendency, the field current must be increased. The amount of field current necessary for normal voltage depends on the load and inductive power factor. In an automatically controlled system, a voltage regulator adjusts the field current to whatever value the load requires. A speed governor is used to maintain generator speed.

The speed governor and the voltage regulator also determine the way an alternator operates on an infinite system, which is a system consisting of several alternators arranged so that the operation of a single generating unit does not affect the voltage or the frequency of the entire system. An alternator connected to an infinite system must operate at rated speed. Its speed governor is then more of a load-regulating device than a speed-regulating device. The amount of the system load that will be carried by the alternator can be controlled by changing the governor setting.

An alternator connected to an infinite system must also operate at rated voltage. The voltage regulator adjusts the field current to control the alternator power factor. The system must, of course, supply the average power factor of the load connected to it. The total reactive kilovolt-amperes on the system may be divided among the various alternators by adjusting their field currents. The voltage regulator then becomes a power-factor regulator for its associated alternator.

Power factor is the ratio of the true power or watts to the apparent power or volt-amperes and is expressed as a decimal or in percentage. Therefore, the equation used to determine power factor of a circuit is

$$PF = \frac{kW}{kVA}$$

To better understand this equation, a knowledge of *true power, inductive reactance*, and *capacitive reactance* is required. An analogy will enhance understanding. For example, imagine a farm wagon being pulled by three farm tractors, as shown in Fig. 3-5. The tractor in the middle is pulling straight forward and may be compared to true power because all its efforts are in the direction that the work should be done. The tractor on the right (#2), however, is not contributing any effort to the task since it is pulling at a 90° angle to tractor #1. This tractor is similar to inductance. The third tractor, the one on the right, is 180° out of phase with tractor #2—or a right angle from tractor #1—and contributes nothing to the desired forward motion; this tractor may be compared to capacitive reactance.

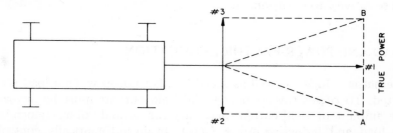

FIGURE 3-5 Wagon being pulled by farm tractor, illustrating how the power factor of a circuit is determined.

If only tractors #1 and #2 were pulling the load, the wagon would go in the direction of the dotted line A. Note that the length of that line is greater than the line to #1. The direction and length of dotted line A might well be called *apparent power* and happens to be the hypotenuse (or diagonal) of a right triangle.

If only tractors #1 and #3 were pulling, then the two would pull the load as indicated by dotted line B, and the length of that line would also be apparent power. If all three tractors were pulling the load, #2 and #3 would cancel one another, and the only useful tractor would be #1.

In an ac circuit, there are always three forces working in varying length. Inductance, for example, is always present in every magnetic circuit and always works in a 90° angle with true power.

In giving the power factor of a circuit, it should be stated whether it is leading or lagging. The current is always taken with respect to the voltage. A power factor of .75 lagging means that the current lags the voltage. The power factor may have a value anywhere between 0 and 1.0 but can never be greater than 1.0.

The power factor in a noninductive circuit (one containing resistance only) is always 1 or 100%. In other words, the product of volts and amperes in such a circuit gives true power.

The speed governor, for most alternators that operate in an infinite system, controls the alternator power output by controlling the prime-mover output. The voltage regulator, by controlling the alternator field current, also controls the voltage and power factor at which the alternator will operate for a given power output.

The alternator must be able to supply the reactive power as well as the real power required by the load. A resistance load, such as an electric incandescent lamp, requires no reactive power and operates at unity power factor, but a motor which needs magnetizing current requires inductive reactive power and operates at a lagging power factor. The power factor of the combined load is the power factor considered in alternator operation.

TYPES OF ALTERNATORS

The main types of alternators in use today are engine-driven alternators, hydraulic-turbine alternators, and steam-turbine alternators.

Most engine-driven alternators are of the salient-pole type and are usually horizontal-shaft machines. Many are made with bolted poles, and damper windings are used to offset the tendency toward pulsations in speed imparted by the engine. Ballast rings or flywheels are also provided on some machines to give them extra weight and inertia to further smooth out the speed fluctuations.

Before covering other types of generators, however, a review of generator characteristics and principles is in order. Therefore, the following paragraphs are designed to give the reader a basic and thorough understanding of ac generators in general.

GENERATOR TERMS

MAGNETISM It is a well-known fact that a magnetic field consisting of concentric circles is created around a wire that is carrying electric current. If the current-carrying wire is wound in a coil, a magnetic field is created with a north or N pole at one end and a south or S at the other end. See Fig. 3-6.

ELECTROMAGNETISM Electromagnetism is the lines of force (magnetic field or flux) produced around a straight wire or coil carrying an electric current. The unit of measure is amperes per minutes.

ELECTROMAGNETIC INDUCTION This term is the producing of an electric current in a conductor when it is moved in a magnetic field so as to cut the lines of magnetic force.
Another application of the principle is found in an alternating-current generator where the magnetic field is made to cut across stationary conductors in order to produce voltage and current.

MAGNETIC SCREEN This is a material (usually iron) used for enclosing certain measuring instruments to shield them from the effects of external stray magnetic fields.

MAGNETIC INDUCTION This occurs when an iron bar is placed in a magnetic field so the lines of force passing through the bar will magnetize it. See Fig. 3-7.

SELF-INDUCTANCE Self-inductance is an electromagnetic induction of a voltage in a current-carrying wire (inductor) when the current in the wire itself is changing with the increasing and decreasing magnetic field around the wire. The polarity of an induced voltage opposes the change in current that produced it and attempts to prevent the current from increasing. See Fig. 3-8.

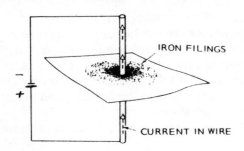

IRON FILINGS

CURRENT IN WIRE   FIGURE 3-6 Electromagnetism.

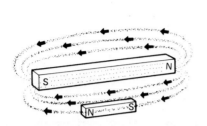

FIGURE 3-7 Magnetic induction.

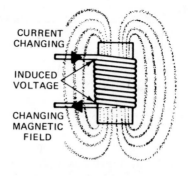

FIGURE 3-8 Induced voltage.

MUTUAL INDUCTANCE The mutual inductance between two coils in a circuit is the quotient of the flux linkage produced in one coil divided by the current, in another coil, which induces the flux linkage.

ELECTRICAL GENERATOR An electrical generator or alternator is a machine so constructed that a voltage is generated when its rotor is driven by an engine or other prime mover.

ALTERNATOR Some alternators have a revolving armature, and others have a revolving field that produces alternating current. In battery-charging alternators, the generated ac is internally changed to direct current (dc) or is rectified before it reaches the output terminals.

ARMATURE This is the rotating coil in a revolving armature generator. The armature in a revolving field generator is stationary.

FIELD The field is the stationary field coils and poles in a revolving armature generator or the rotating field coils and poles in a revolving field generator or alternator.

REVOLVING ARMATURE GENERATOR This is a two- or four-pole generator in which power is generated in the revolving coil and delivered to the load via a commutator or slip rings and brushes.

REVOLVING FIELD GENERATOR This is a two- or four-pole generator in which power is generated in the stationary coil while the magnetic field rotates within it. Excitation currents to the field pass through brushes and slip rings.

COMMON GENERATOR VOLTAGES These are 120 and 240 single phase and 120/208, 480, and 600 three phrase.

COMMUTATING POLE (INTERPOLE) GENERATOR This is used to compensate for the magnetic field distortion and to prolong

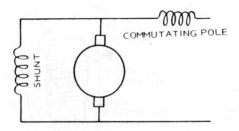

FIGURE 3-9 Commutating pole.

brush life and provide good commutation without serious arcing on large dc generators like the 6 CCK arc welders. One commutating pole is used for every two field poles, as shown in Fig. 3-9.

CYCLE One cycle or hertz is a complete period of flow of alternating current in both directions. One cycle represents 360° rotation of an armature.

HERTZ This is a unit of frequency (Hz), one cycle per second, written as 50-Hz or 60-Hz alternating current, and so on.

FREQUENCY Frequency of alternating current is the number of cycles per second; 60-Hz alternating current makes 60 complete cycles of flow back and forth (120 alternations) per second. A conventional alternator has an even number of field poles arranged in alternate north and south polarities.

Current flows in one direction in an ac armature conductor while the conductor is passing a north pole and in the other direction while passing a south pole. The conductor passes two poles during each cycle. A frequency of 60 Hz requires the conductor to pass 120 poles/sec.

In a four-pole alternator, the equivalent speed would be 30 rev/sec or 1800 rev/min:

$$\text{frequency (Hz)} = \frac{P \text{ (number of poles)}}{120} \times \text{rpm}$$

FREQUENCY METER This indicates frequency in cycles per second (hertz), which is proportional to the engine speed.

EXCITER An exciter is a generator which supplies current to excite or magnetize the fields of an alternator. An exciter may be a separate machine or be combined with the alternator, or it can be any device that supplies excitation currents to an alternator.

Exciters are necessary for the generators with speeds through 1800 rpm and are usually direct-connected; the armature is mounted to the generator shaft extension. See Fig. 3-10.

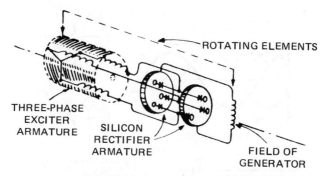

THREE-PHASE ALTERNATING-CURRENT
GENERATED IN EXCITER ARMATURE WINDING

FIGURE 3-10   Exciter circuit. (Courtesy Onan.)

SINGLE PHASE   A single-phase, alternating-current system has a single voltage in which voltage reversals occur at the same time and are of the same alternating polarity throughout the system. See Fig. 3-11.

THREE PHASE   A three-phase, alternating-current system has three individual circuits or phases. Each phase is timed so the current alternations of the first phase is $\frac{1}{3}$ cycle (120°) ahead of the second and $\frac{2}{3}$ cycle (240°) ahead of the third [Fig. 3-12(a)]:

three-phase voltage = one-phase voltage × 1.732

STAR CONNECTION   The Y (wye) or star connection has one end of each coil connected together as at 0 [Fig. 3-12(b)]. If V represents voltage generated in one conductor (or all the series conductors in one phase), it will also be equal to the voltage from the common point 0 (commonly called star or neutral) to terminal A, C, or B. The voltage between terminals A to C, C to B, or B to A will be 1.73 × V.

DELTA CONNECTION   If the delta connection is used, the conductor ends are connected as in Fig. 3-12(c). In that case the voltage in each coil is the voltage between terminals.

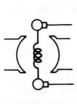

ONE-PHASE

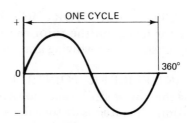

FIGURE 3-11   Single-phase system. (Courtesy Onan.)

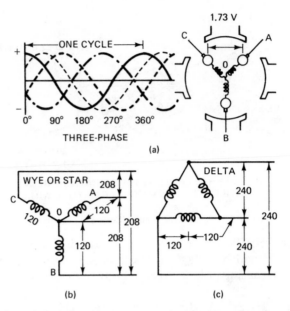

FIGURE 3-12 Three-phase system. (Courtesy Onan.)

COMMUTATOR This is the radial copper segments on the rotor of an electric generator or motor. They conduct current from the rotating windings to the brushes, or vice versa for excitation.

BRUSHES These are the carbon or copper spring-loaded sliding contacts that bear on the commutator or slip rings and carry current to or from the rotating part of the generator.

BRUSH POSITION The best brush position is usually halfway between the N and S poles where no magnetic lines of force are cut by the coil and no voltage is generated.

MAGNETIC NEUTRAL This is the point halfway between the N and S poles where no voltage is generated.

SLIP (COLLECTOR) RINGS These are smooth (insulated) conductor bands on the rotor shaft used to feed ac from the armature to the brushes in a revolving armature generator.

ANTIFLICKER POINTS Because the engine speed increases during the power stroke and slows down on the compression stroke, the waveform generated is distorted, causing flicker in the lights (Fig. 3-13). During the dips in the waveform the flicker resistor is shorted out so more voltage is generated which will average out the dips.

STABILITY The operating of generators is of prime importance.

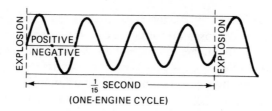

FIGURE 3-13   Flicker. (Courtesy Onan.)

Poor stability becomes evident when sudden changes occur in operating conditions caused by

1.   A short circuit
2.   A ground
3.   Sudden load change
4.   System switching surge
5.   Lightning stroke

Lack of stability prevents units from carrying their load through such sudden changes and may even permit them to drop out of step if operating in parallel.

REGULATION   Regulation is defined as the rise in voltage (field current and speed remaining constant) when full load is thrown off the generator:

$$\text{percent regulation} = \frac{\text{voltage at no load} - \text{voltage at full load}}{\text{voltage at full load}} \times 100$$

Standard voltage regulation figures are

Full-load 80% power factor lagging 40%
Full-load 100% power factor 25%

AIR GAP EFFECT ON FIELD STRENGTH   The effect of an air gap on the total field strength of an electromagnet is very pronounced because air has much higher reluctance than iron. If the width of the air gap is doubled, the reluctance quadruples, and field strength reduces to about half (Fig. 3-14). Reluctance increases as the square of the distance.

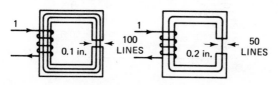

FIGURE 3-14   Air gap effect on field strength. (Courtesy Onan.)

INDUCTANCE Any device with iron in its structure has a magnetic inertia which opposes any change of current or change in voltage; any coil operating in an ac circuit—even with an air core—will have inductance also. Inductance is found only with alternating current since the voltage is continually changing in instantaneous value while the magnetic inertia is causing changes in current to lag behind changes in voltage. The characteristic that causes the magnetic inertia is self-inductance. Inductance like self-inductance is measured in henries (H).

REACTANCE Reactance is the opposition to the change of current or voltage in an ac circuit. The alternating current applied to the circuit will be affected by capacitance or inductance and will cause the current waves to get out of step with the voltage waves.

INDUCTIVE REACTANCE The opposition of self-inductance to current flow is called inductive reactance because it causes the current to lag behind the voltage that produces it. The unit of reactance (X) is the ohm (Ω), the same as for resistance. However, resistance opposes both ac and dc, while reactance opposes only ac. In an ac circuit with a definite frequency, inductance results in an inductive reactance (X_L) which is measured in ohms and is determined by

$$X_L\ (\Omega) = 2 PifL \qquad \text{or} \qquad X_L\ (\Omega) = 2 \times 2.1428 \times \text{Hz} \times \text{H}$$

$$\text{or} \qquad 6.28 \times f \times L$$

Figure 3-15 shows how the inductive reactance of a coil reduces the current to a light bulb:

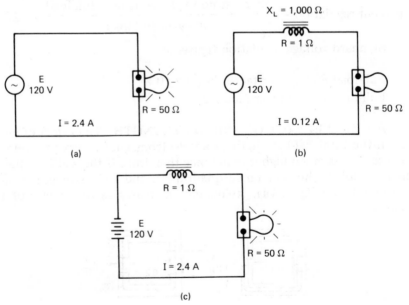

(a)

(b)

(c)

FIGURE 3-15 Inductive reactance. (Courtesy Onan.)

1. In Fig. 3-15(a) with no inductance, the ac voltage source produces 2.4 A to light the bulb.

2. In Fig. 3-15(b), connecting a coil with 1-Ω resistance but having an inductive reactance of 1000 Ω limits the alternating current to .12 A at the bulb, so it cannot light.

3. In Fig. 3-15(c) with the coil still in series with the bulb but applying battery voltage (dc without current variations), the bulb lights as though the coil is a length of wire with 1-Ω resistance.

CAPACITIVE REACTANCE The opposition of capacitance to the change of ac voltage causes the current wave to lead the voltage wave and is called capacitive reactance. Impedance is always a combination of resistance and capacitance and/or inductance and is valid only on ac circuits. The standard symbol for impedance is Z. It is measured in ohms and is determined in ac circuits by

$$X_C = \frac{1}{2 \, Pi f C} \quad \text{or} \quad X_C = \frac{1}{2 \times 3.1428 \times Hz \times C}$$

$$\text{or} \quad \frac{1}{6.28 \times f \times C}$$

1. Pure resistance (R) opposes current flow but allows current to remain in phase with voltage.

2. Inductive reactance (X_L) causes current to lag voltage by 90°.

3. Capacitive reactance (X_C) causes current to lead voltage by 90°.

IMPEDANCE (MΩ) This is the total opposition to current flow in an ac circuit:

$$\text{ac ohms} = \sqrt{R^2 + (X_L - X_C)^2}$$

4

Engine-Driven
and Gas-Turbine Generators

Essential service loads, such as hospitals, most public buildings, and many kinds of industrial operations, that cannot tolerate a power outage require two separate sources of power. These sources, of course, should not be paralleled. Rather, an automatic transfer system is generally used and is built in many different designs and various voltages.

The most important factor in selection of a standby system is, of course, size and/or capacity. A careful study should be made to determine exactly what degree of service it is desirable to maintain during a power outage. How much and what kind of equipment must the system operate? What is the total wattage?

It is advisable at this stage to adopt a long-range view since it often becomes necessary to add more equipment to the load in future years. It may be desirable to provide full protection at the outset, since the higher cost for a larger unit is offset by special circuit wiring costs necessary for selected protection. As needs grow, the set may still be made adequate by restricting protection to essential equipment. Of course, the decision of whether to provide full or selected protection will most often be decided by the particular operation for which the unit is provided. A hospital, for example, requires full protection for all essential equipment, while another building may only wish to maintain emergency lighting.

Another factor to consider at this stage is the altitude of the locale and the effect it may have on the rating of the standby set. If the point of installation is at a higher altitude than that of the manufacturer, the unit is generally derated 4%/1000 ft above sea level.

Also, complex factors, such as surge currents of electric motors, the magnetizing current of transformers, feedback from elevator loads, heating effects, and electromagnetic interference from solid-state loads, must be thoroughly explored and understood before system parameters are finalized.

STANDBY SYSTEM PARAMETERS

FUEL Before the engine can be selected, careful consideration must be given to selecting the proper type of fuel. Often, the type of fuel is one of the most important considerations in final engine selection.

Fuel options are gasoline, liquid propane (LP) gas, natural gas, and diesel. Specific advantages of each are discussed later in this chapter. However, several general points which may outrule all others should be considered at the outset:

1. Availability of the various fuels in your particular location

2. Local regulations governing the storage of gasoline and gaseous fuels.

COOLING Air and liquid cooling systems are available. Liquid cooling may be either a radiator or freshwater cooling system. The determining factor between air or water cooling will generally be the size of the unit. Air cooling must be confined to smaller units up through 15 kW. In these smaller sizes it is possible to force air over the engine and generator at a sufficient rate to cool it properly and thus take advantage of lower equipment and maintenance costs. Larger units, above 15 kW, should—as a rule—be equipped with a liquid system for maximum cooling protection.

Regardless of what capacity or cooling system is used, location, size, and temperature of the room must be considered. City water cooling, remote radiators, and heat exchangers are available for special applications.

LOAD TRANSFER SWITCH Electric power companies require transfer switches and usually supervise their installation, since power lines on which utility personnel work are affected.

There are two types of transfer switches: manual and automatic. The manual transfer switch must be operated by a person on duty either at the plant or a remote station. The plant is started by a manual switch, and the load is transferred from one source to another by a hand-operated double-throw switch.

An automatic transfer switch starts the plant and transfers the load automatically, not requiring the attention of an operator. With automatic transfer switches, the power outage can be limited to less than 10 sec. An automatic load transfer switch is almost a necessity for applications where uninterrupted power is of prime importance, such as a hospital, or where equipment is remotely located or where health or safety is at stake.

BASIC GENERATORS

Generators and alternators used for emergency standby power are usually of two basic types: dc and ac revolving armature and ac revolving field. The output voltage of these generators—depending on the type—may be regulated by field pole saturation, self-excitation, rectifier excitation, or brushless exciter currents. In any case, residual voltage and dc excitation are required for voltage buildup and output voltage regulation.

The basic generators discussed in this chapter have an excitation field winding but differ otherwise because of the purpose for which they are intended. Variations include special-purpose generators and compound generators designed for specific types of electrical loads.

An armature is the rotating part of a magnetic field. In revolving armature generators, the power-producing part is the armature; the field power is stationary. In revolving field generators, the field power rotates and is called the rotor; the power-producing part of this generator is stationary. See Fig. 4-1 and Table 4-1.

TABLE 4-1 Generator Comparison Factors

Factors	Revolving Armature	Revolving Field
Rating	300 W to 10 kW	1.2 kW to 750 kW
Type of regulation	Inherent or external	External
Percent of regulation	± 3 to $\pm 7\%$	± 2 to $\pm 3\%$
Output voltage	480, maximum	600
Slip rings and brushes	Limit output power	Slip rings—1.2 to 6.0 kW No slip rings—6.0 to 750 kW
Construction	Cheaper to build	Static exciter or solid-state control
Rotating part	Armature	Rotor
Stationary part	Frame assembly	Stator
Electric power source	Armature	Stator
Excitation source	Frame assembly	Rotor
NEMA output leads	M_1, M_2, M_3, M_4	T_0, T_1, T_2, T_3, etc.
Components		
Armature shaft, stack, bearings, windings, commutator, rings Frame assembly, frame mounting, feet, pole shoes, field coils, and brush rig		Stator laminations, welded bar, windings, and leads Rotor shaft, stack, slip rings, windings, bearings, drive disk

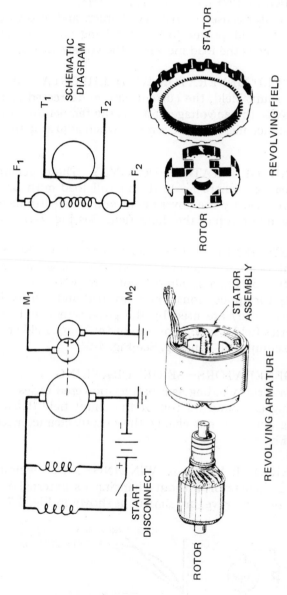

SCHEMATIC DIAGRAM

T_1

T_2

F_1

F_2

M_1

M_2

START DISCONNECT

REVOLVING ARMATURE

ROTOR

STATOR ASSEMBLY

STATOR

REVOLVING FIELD

ROTOR

FIGURE 4-1 Generator rotor and stator laminations. (Courtesy Onan.)

43

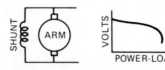

FIGURE 4-2 Shunt field generator with commutator. (Courtesy Onan.)

SHUNT FIELD GENERATOR WITH COMMUTATOR This popular type of generator is used on battery chargers and revolving exciters and for supplying field power for ac revolving armature generators. Note in Fig. 4-2 that as the load increases, the voltage drops.

SHUNT FIELD GENERATOR WITH RHEOSTAT By adding a rheostat to the shunt field, the current can be decreased in the shunt field and also lower the ac voltage ±5%. As the temperature of the wire goes up, the resistance increases the same as when adjusting the rheostat. See Fig. 4-3.

COMPOUND GENERATOR—DC ONLY These generators are made for applications where no load and full load must be the same voltage. They are used by the government and on electromagnets. The series field is wound on top of the shunt field. See Fig. 4-4.

COMPOUND GENERATORS—NONCRANKING By combining series and shunt fields, the generator will produce voltage that rises, decreases slightly, or remains almost constant with load (dc only). This generator is for applications where no load and full load must be the same voltage. They are used by the government and on electromagnets. The series field is wound on top of the shunt field and is sometimes called the compound winding. See Fig. 4-5.

SHUNT GENERATORS—SERIES CRANKING Cranking-wound generators use a series winding as a battery-powered series motor. The cranking winding is wound on top of the shunt field. It is only used during cranking and does not change the output characteristics of the generator. See Fig. 4-6.

DC GENERATOR WITH AC WINDING An ac winding can be added to any dc generator without changing its external appearance. Both ac and dc output curves are similar, as shown in Fig. 4-7.

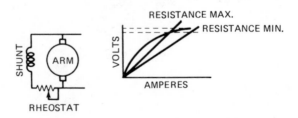

FIGURE 4-3 Shunt field generator with rheostat.

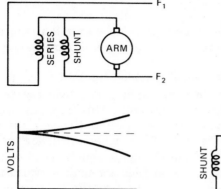

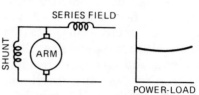

FIGURE 4-4 Compound generator — dc only. (Courtesy Onan.)

FIGURE 4-5 Compound generator — noncranking. (Courtesy Onan.)

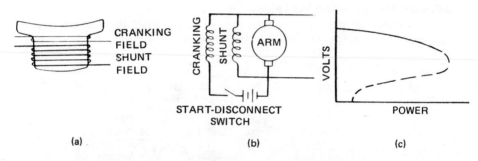

(a)

(b)

(c)

FIGURE 4-6 Shunt generator — series cranking. (Courtesy Onan.)

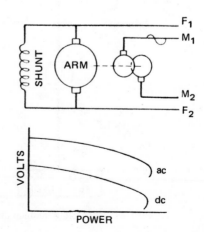

FIGURE 4-7 Generator with dc winding. (Courtesy Onan.)

GENERATOR VOLTAGE BUILDUP

Before a generator will produce voltage, it is necessary that the exciter currents be produced. The steps of voltage buildup are shown in Fig. 4-8.

The difference in voltage developed between the two ends of the armature loop when it is rotated through the magnetic field causes an excitation current to flow through the field coils. This flow of current through the field coils, which are wrapped around the pole pieces, makes the poles electromagnets (Fig. 4-9). The magnetic field developed by these electromagnets strengthens the *residual* magnetic field of the pole pieces and greatly increases the total field strength between the poles. The rotating armature loop, cutting through the stronger magnetic field, increases the induced voltage in the loop, which, in turn forces more current through the field coils. This creates an even stronger magnetic field with more lines of force. The armature cutting through this stronger magnetic field develops still more voltage. In this manner, the voltage of a generator is built up. This voltage increase continues rapidly until the rated value of the generator is reached or if not regulated builds up to a higher value.

A basic dc generator schematic diagram with the exciter field is shown in Fig. 4-10. The voltage between the two ends of the armature

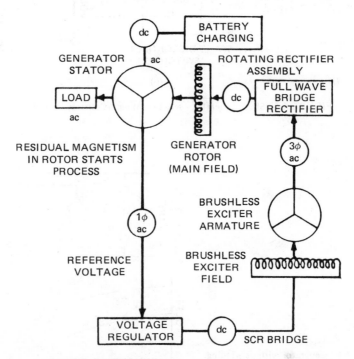

FIGURE 4-8 Voltage buildup diagram. (Courtesy Onan.)

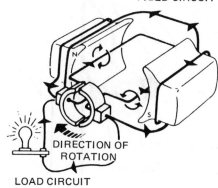

FIELD CIRCUIT

DIRECTION OF
ROTATION

LOAD CIRCUIT

FIGURE 4-9   Field and load circuits.
(Courtesy Onan.)

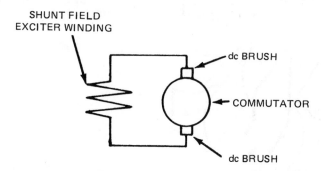

SHUNT FIELD
EXCITER WINDING

dc BRUSH

COMMUTATOR

dc BRUSH

FIGURE 4-10   dc generator with shunt field excitation diagram. (Courtesy
Onan.)

loop causes a flow of current through the load circuit as indicated when
a light bulb is placed in the load circuit (Fig. 4-9).

The ends of the armature loop are securely attached to a split
ring called a *commutator* (Fig. 4-11). Riding on the commutator are the
brushes. It is from these brushes that the voltage and the resulting current are transmitted to both the field and the load circuits. This is called
commutation.

When located in a vertical position, the armature loop cuts no
magnetic lines of force. Therefore, no voltage is generated at this point.
When the armature is in this position, the split in the commutator ring
is passing under the brushes. The split in the commutator ring makes
the direction of voltage and current the same with respect to the
brushes. Since no voltage is present in the loop at this particular instant,
there is no current flow. Consequently, no arc or spark occurs when
one commutator bar moves from under the brush and the adjacent bar
replaces it. This halfway point is called the *neutral point* (Fig. 4-12).

The magnitude and direction of voltage and the neutral points
during one complete revolution (cycle) of the armature are shown in
Fig. 4-13.

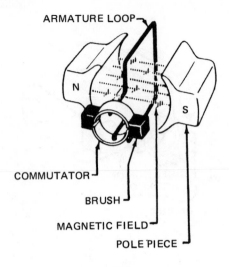

ARMATURE LOOP

N

S

COMMUTATOR

BRUSH

MAGNETIC FIELD

POLE PIECE

FIGURE 4-11 Basic components.
(Courtesy Onan.)

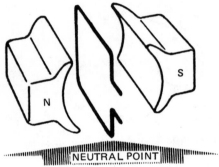

N

S

NEUTRAL POINT

FIGURE 4-12 Neutral point. (Courtesy
Onan.)

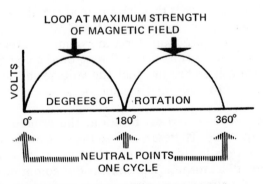

LOOP AT MAXIMUM STRENGTH
OF MAGNETIC FIELD

VOLTS

DEGREES OF   ROTATION

0°   180°   360°

NEUTRAL POINTS
ONE CYCLE

FIGURE 4-13 One cycle. (Courtesy Onan.)

Just as electrical circuits must be complete, there must also be a complete magnetic circuit. Figure 4-14 shows the magnetic circuit of a two-pole generator. It is important to remember that all air gaps in the magnetic circuit act as high reluctance and cut down the strength and effectiveness of the magnetic field.

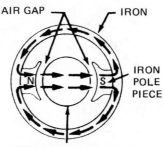

IRON
POLE
PIECE

FIGURE 4-14   Magnetic circuit. (Courtesy Onan.)   IRON ARMATURE CORE

For a more realistic picture of an actual generator, see Fig. 4-15, which shows an armature containing additional loops of wire embedded in the slots of an iron core. All loops are connected together at the commutator. Any voltage developed in one loop is added to the voltage developed in the other loops since they are connected in *series*.

Voltage is increased by adding conductors to the armature, as shown in Fig. 4-16. When more conductors and more commutator bars are used, the voltage waves overlap, producing an almost constant value of dc voltage at a given speed and load.

Three fundamental and interrelated factors are always necessary for developing current and voltage or electrical power from a generator:

1. Strength of magnetic field
2. Number of conductors on armature
3. Speed of armature rotation

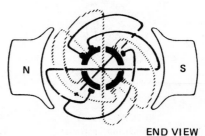

FIGURE 4-15   Series-connected loops.
(Courtesy Onan.)

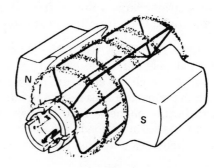

FIGURE 4-16   Multiconductors. (Courtesy Onan.)   END VIEW

Generator Voltage Buildup   49

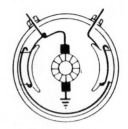

FIGURE 4-17 Electrical diagram.   FIGURE 4-18 Eddy currents.

Figure 4-17 shows a simplified electrical diagram commonly used to indicate brush-type generator wiring arrangements.

HEAT The heat produced by current flow during the operation of generators is carried away by a blower which draws air through the generator. This permits greater output without overheating the unit. Ventilated generators carry a higher current rating than nonventilated units of the same size.

EDDY CURRENTS Another source of heat, called *iron loss*, is present in the armatures of all generators. The iron core of the armature acts as one large conductor which cuts magnetic lines of force as it revolves, generating voltage within the core itself. This action results in current flow called *eddy currents* (Fig. 4-18). These currents produce heat which is added to the heat developed by current flow in the conductors.

To reduce the effects of eddy currents as much as possible, the iron core of the armature is laminated. See Fig. 4-19. The thin laminated sections prevent large voltages from developing, and the eddy currents are kept small. Therefore, less heat is developed in the armature.

ARMATURE REACTION Armature reaction is the distortion of the normal magnetic field caused by a separate magnetic field set up in the iron on which the armature conductors are wound. Current flowing through the armature coils results in another magnetic field that opposes or distorts the magnetic field between the N and S poles, as shown in Fig. 4-20. As variable loads are applied to the smaller generators, these magnetic fields react against the excitation field, proportionally creating

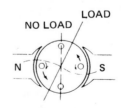

FIGURE 4-19 Typical lamination sections.

FIGURE 4-20 Distorted magnetic field. (Courtesy Onan.)

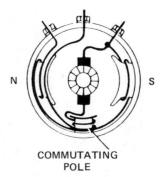

FIGURE 4-21 Commutating pole generator schematic. (Courtesy Onan.)

COMMUTATING
POLE

additional horsepower demands on the engine. The results are drops in engine speed and drops in the generator output voltage and frequency. Mechanical governors on the small units are unable to fully correct for the decreases in speeds caused by the larger load demands.

Remember, all load currents flow through the conductors of the armature, so the greater the current flow, the greater the strength of the magnetic field and the greater the countertorque.

COMMUTATING (INTERPOLE) GENERATORS Commutating pole generators use an extra pole to neutralize the harmful voltages generated by the distorted magnetic field at the neutral points and thus prevent arcing of the brushes. The commutating pole is a narrow pole piece placed halfway between the N and S poles, illustrated in Fig. 4-21.

REVOLVING ARMATURE GENERATORS

In a revolving armature generator the rotating coil is the armature; the magnet is the field. If two magnets are used (four poles), the unit is a four-pole generator. The ac output frequency doubles that of a two-pole unit operating at the same speed.

The standard ac-dc revolving armature generator is shown in Fig. 4-22 and is either a two- or four-pole unit, is self-excited, and is inherently regulated. The armature has both ac and dc windings in it.

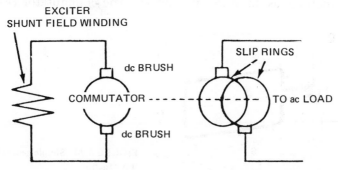

FIGURE 4-22 Revolving armature generator. (Courtesy Onan.)

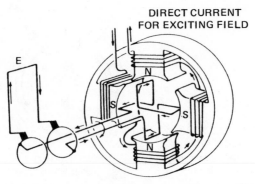

FIGURE 4-23   Basic four-pole ac generator. (Courtesy Onan.)

PRODUCING AN ELECTROMAGNETIC FIELD   Electromagnetism produces variable magnetic fields required in most generator applications. Field strength is accomplished with electromagnets, which are field windings with an iron core, as shown in Fig. 4-23. The magnet field strength is proportional to the current in the field. The more current in the field, the more magnetic lines of force and the higher the output voltage. Direct current for field excitation can be supplied either from an external source, such as batteries, or from the generator's output, known as self-excitation.

There are four basic self-excited generator configurations:

1.   A shunt field.
2.   A shunt field and series cranking.
3.   A series field.
4.   A compound shunt and series field depending on the generator application and the voltage control method. The field current determines the output voltage, which is regulated by changing the field currents.

USING AN AC GENERATOR TO PRODUCE DC   Converting the ac generator to a dc generator requires the addition of a commutator which functions as an automatic switch. The commutator (Fig. 4-24)

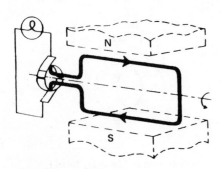

FIGURE 4-24   Simple generator. (Courtesy Onan.)

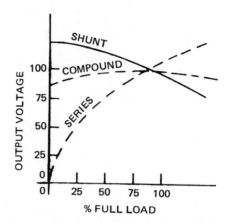

FIGURE 4-25 Voltage regulation of shunt, series, and compound field windings. (Courtesy Onan.)

inverts half of the output voltage by reversing the relationship of the armature coils to the output wires each time the voltage is zero. This produces pulsating direct current in the output leads of the generator.

In shunt-wound generators, the field current depends on the armature output voltage. If a heavy load reduces voltage slightly, the shunt field current drops, and the resulting output drops further. This results in a voltage drop with an increased load (Fig. 4-25).

With a series winding, all load currents pass through the field, so the output voltage is almost zero with no load and increases as the load increases. With both series and shunt fields combined, the generator produces a voltage characteristic that rises, decreases, or remains almost constant with the load. A combined series and shunt field is known as a compound-wound field.

AC GENERATOR EXCITATION   The ac generator has requirements different from dc generators. To develop direct current for its field, an additional winding and commutator, known as the exciter, are required on the armature. The voltage regulation method also differs. All ac revolving armature generators are inherently regulated. Late-model small generators use heavy-duty silicon rectifiers for ac to dc conversion to produce field winding excitation. Two slip rings and brushes transfer the current to the field winding.

Further understanding of the generator requires knowledge of phenomena of iron when magnetized. First, iron is saturable: Increasing the current in a coil around a piece of iron will increase the magnetic flux in the iron only to a point. Above that point, increased current has a reduced effect until finally further increases do not change the field strength. At this point, the magnet is said to be saturated. Second, when current is stopped, the magnet retains some of its strength or has residual magnetism (Fig. 4-26). As the current in the coil is increased, the magnetic strength finally reaches saturation point A. When the current is reduced to zero, the strength stays at point B.

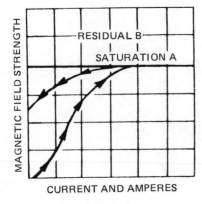

FIGURE 4-26 Saturation and residual magnetism. (Courtesy Onan.)

When the generator supplies its own field, an initial jolt is required to start voltage buildup. The residual magnetism supplies this by maintaining a small magnetic field when the generator is stopped. Once the armature is rotating, this residual field induces a small voltage in the armature which is fed back through the field windings to reinforce the field. The reinforced field induces a larger voltage which further reinforces the field. Buildup continues until limited by the generator characteristics or is regulated by the voltage control.

The maximum voltage a generator will produce is determined by the field saturation. At this point, further increases in field current will not change the field strength and so will not affect the generator's voltage. This effect is utilized in ac revolving armature generators and is called inherent regulation. Inherent regulation through saturation is not used in most dc generators. The operator controls the output voltage.

VOLTAGE REGULATION   The output of a generator must be regulated or controlled. Voltage control is needed to protect not only the generator but also the load electrical system. Without voltage control, light bulbs, external wiring, relay coils, contact points, and all the other electrical components would be endangered by the high voltage, resulting in short life or damaged accessories.

A voltage regulator indirectly controls the output of a generator by controlling the exciter shunt field currents. These devices are variable resistances which are used to increase or decrease the exciter currents to obtain a change in generator output voltage.

Voltage regulators provide the corrective action (rise in voltage) required to offset the armature demagnetizing effect and overcome the voltage dip which occurs when a load is thrown on the generator. The voltage regulator provides increased excitation whenever the voltage dips beyond the generator's sensitivity limits, sensitivity being a narrow band within which no regulation is required.

The maximum dip in voltage is usually not affected by the amount of initial load on the generator, but the initial load does affect the voltage level to which the generator will recover after a dip occurs.

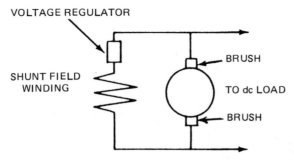

FIGURE 4-27   Control of shunt field. (Courtesy Onan.)

VOLTAGE REGULATORS   Most dc generators are self-excited; that is, the generator supplies its own excitation currents. The shunt field (excited field) is connected directly to the dc (commutator) brushes. Thus, the dc armature produces both the field currents and the load currents. The output voltage is controlled through the use of a voltage regulator (Fig. 4-27). The regulated-type generators employ some type of automatic voltage regulator or manually operated rheostat to control field currents.

REVOLVING FIELD GENERATORS

Instead of turning a coil of wire in the magnetic field, the wire coil is fixed and the field is rotated inside the coil. The result is a revolving field generator (Fig. 4-28) which produces only ac and is commonly called an alternator. The revolving field is the rotor, and the stationary fixed windings are the stator. With this type of generator there is no need for slip rings to transfer power from the stationary (armature) coils.

Slip rings are used only to supply direct current to the electromagnetic field in the rotor. Some generators use a static exciter and a nonrotating device that converts one phase of the ac output to dc and regulates current to the field. This static exciter is called a magneciter (Fig. 4-29).

The output frequency of the revolving field generator depends directly on its rotating speed. The voltage output is determined by the rotating speed, the number of turns in the stator, and the strength of the controlling field currents. The magneciter allows for output voltage adjustment in a limited range—3% at a steady speed—and has rapid recovery capabilities from a sudden load application or removal.

Some older dc generators use a series shunt field connection for the exciter field (Fig. 4-30).

The exciter circuit of a revolving field generator is shown in Fig. 4-31. The direction of current flow is indicated by an arrow. The exciter

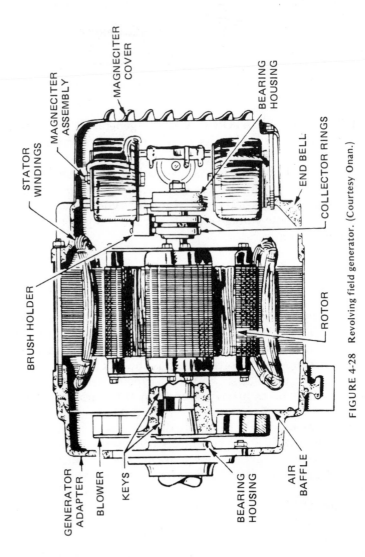

STATOR WINDINGS

MAGNECITER ASSEMBLY

MAGNECITER COVER

BEARING HOUSING

BRUSH HOLDER

END BELL

COLLECTOR RINGS

ROTOR

GENERATOR ADAPTER

BLOWER

KEYS

BEARING HOUSING

AIR BAFFLE

FIGURE 4-28 Revolving field generator. (Courtesy Onan.)

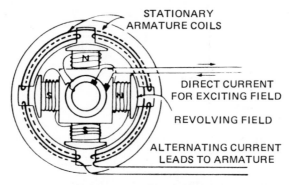

DIRECT CURRENT
FOR EXCITING FIELD

REVOLVING FIELD

ALTERNATING CURRENT
LEADS TO ARMATURE

FIGURE 4-29   Revolving field generator (magneciter). (Courtesy Onan.)

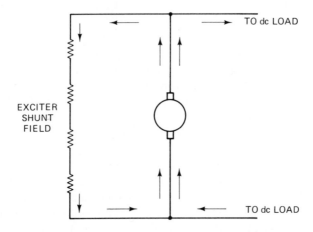

TO dc LOAD

EXCITER
SHUNT
FIELD

TO dc LOAD

FIGURE 4-30   Series exciter field. (Courtesy Onan.)

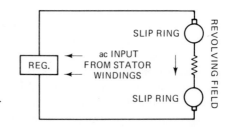

SLIP RING

ac INPUT
REG.   FROM STATOR
WINDINGS

REVOLVING FIELD

SLIP RING

FIGURE 4-31   Series-connected exciter
field. (Courtesy Onan.)

field is shown connected in series. The regulator measures the ac output voltage and controls the field current automatically to maintain a nearly constant output voltage.

RECTIFIER EXCITED CIRCUIT Portable lightweight model generators employ a simple rectifier circuit to provide dc to the exciter shunt field as well as providing ac to the generator load (Fig. 4-3). When a positive (+) portion of the current is picked up from the slip

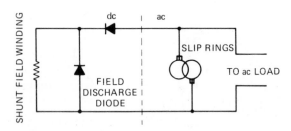

FIGURE 4-32 Rectifier excited circuit. (Courtesy Onan.)

ring, the forward-biased diode rectifies the ac to dc half-wave current and directs the current through the shunt windings and through the other slip ring (–) to the armature coils. At the same time, the positive (+) half-cycle portion of the ac is directed through the ac load and returns to the other side of the armature windings.

On the next half of the cycle, the negative (–) portion of the ac is routed from the slip rings to the ac load. The reverse-biased diode effectively blocks current from the shunt field during this half of the cycle. The other diode across the shunt field circuit acts as a field discharge circuit to prevent high currents from being produced by the collapse of the magnetic field around the shunt winding.

BRUSHLESS GENERATOR AND REGULATOR   All the UR, UV, and YB and YD series generators, as well as other newer models, are brushless exciter controlled. Alternating current from one stator output winding is fed through a normally closed field breaker to a regulator bridge on the voltage regulator chassis (Fig. 4-33).

The full-wave bridge is composed of two silicon controlled rectifiers (SCRs) and three rectifier diodes. This SCR bridge delivers rectified current from the voltage-sensing circuit to the generator's exciter field in timed pulses, determined by an increase in gate voltage, to change positive and negative (*PN*) junctions from reverse to forward bias. This change lets current flow until the half cycle reaches zero voltage. The moment of gate energizing can be set to fire early or late in the half-wave (dc), energizing shifts to an earlier point when the voltage-sensing circuit detects a drop in voltage output. Firing earlier causes the exciter to produce greater output and thus increase generator output.

The generator output increases because the exciter field causes a rise in the exciter armature output which is converted to dc by the rotating diode network (mounted on the exciter armature) and supplied to the generator rotor field. A rise in generator output then causes the voltage-sensing circuit in the regulator to fire the SCRs later in the ac wave and reduce exciter output to a normal level.

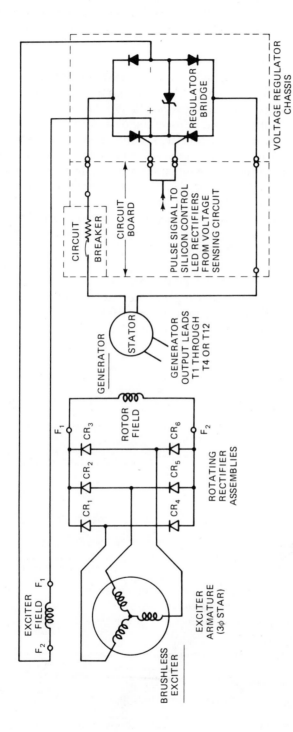

FIGURE 4-33 Brushless generator control circuit. (Courtesy Onan.)

VOLTAGE WAVE SHAPES   The voltage wave shapes in the gating and power stage circuits from the generator stator to the exciter field are shown in Fig. 4-34.

MAGNECITER EXCITATION   All revolving field alternators require dc excitation for their field coils. The basic difference between the magneciter controlled generator and a standard revolving field generator lies in the excitation and voltage regulation circuits.

With static exciter (magneciter) controlled units, excitation is provided by taking part of the generator ac output voltage and rectifying it to dc as required for excitation current. The dc is then fed to the field coils through a set of slip rings and brushes and controlled within the desired limits by the voltage regulator.

Major components of the magneciter circuit are the rectifier power supply, the control, and the voltage regulator, which provide excitation, control, and regulation.

EXCITER FIELD EXCITATION   Excitation is accomplished by a silicon bridge rectifier power supply. The excitation current is taken from the ac output of the generator, rectified, and fed through slip rings and brushes to the revolving field of the generator. The exciter

GENERATOR OUTPUT
WINDING T$_7$ AND T$_8$

CHARGING VOLTAGE ACROSS
CAPACITOR (C$_4$) ON CIRCUIT BOARD.
SLOPE, OR TIME OF BUILD UP VARIES.

PULSE SIGNAL TO SCRs. PEAK
VALUE IS FIXED BUT POSITION
IS VARIABLE

ZENER (Z$_2$) OUTPUT VOLTAGE,
PROVIDES CONSTANT VOLTAGE
ACROSS GATE FIRING ELEMENTS.

G = POINT IN TIME

VOLTAGE TO GENERATOR EXCITER
FIELD FROM SCRs. POINT OF FIRING,
G, VARIES WITH THE EXCITATION
REQUIREMENTS

NOTE: COMPONENTS IN PARENTHESIS ARE SHOWN ON THE UR SCHEMATIC IN 900-150.

THREE-PHASE ALTERNATING
CURRENT GENERATED IN
EXCITER ARMATURE WINDING

RECTIFIED ac (DIRECT CURRENT)
TO ROTOR FIELD WINDING OF THE
ac GENERATOR

RECTIFIER DIODES

FIGURE 4-34   ac to dc conversion wave shapes. (Courtesy Onan.)

+

C

−

ALTERNATOR OUTPUT BRIDGE
RECTIFIED (INVERTED)

FIGURE 4-35   Full-wave rectifier
power. (Courtesy Onan.)

power supply circuit provides full-wave direct current to the generator field windings, shown in Fig. 4-35.

One half-wave is rectified through two diodes, shown in Fig. 4-36. The other half-wave is rectified by the other two diodes to produce full-wave dc field excitation.

AMPLIFIER CIRCUIT   The control component of the magneciter generator controls the amount of current supplied to the field. Control is accomplished by the magnetic amplifier circuit, which is a full-wave bridge rectifier with two saturable (gate) reactors and one control reactor added (Fig. 4-37). The saturable and control reactors are toroid coils.

TOROID COIL   A toroid coil is an electromagnet without an air gap; it closes the magnetic loop and concentrates the magnetic lines of force within the core.

SATURABLE REACTORS   Each saturable reactor (one for each half-wave) consists of a toroid core with an output winding and a control winding. The output winding connects to the rectifier circuit, and the control winding connects to the regulator component.

1.   Figure 4-38 shows how the output winding of the saturable reactor is connected into the exciter circuit for half-wave control.
2.   Figure 4-39 shows how the control windings or both saturable reactors are connected to complete the magnetic amplifier circuit.

The characteristics of a saturable reactor core are such that once magnetized to saturation by the current in its output windings, it is permanently magnetized and remains so even after the current ceases to flow in those windings.

A rectifier in series with the output windings of the saturable reactor allows current flow in only one direction. Thus, the line voltage acts only to magnetize (maintain saturation) but cannot demagnetize the core. Demagnetization is necessary and is handled by the control windings of the saturable reactor and a regulator.

When the saturable reactor cores are in a saturated condition, the reactors are unable to oppose the line voltage. Thus, the full rectified line voltage is applied to the alternator field winding. Uncontrolled, this high excitation current would result in alternator output voltages

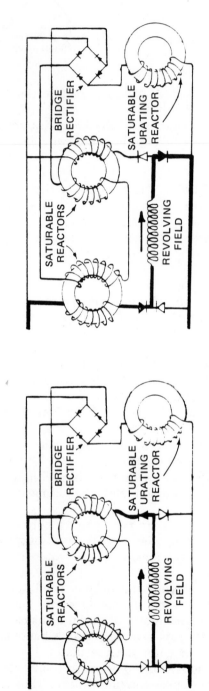

FIGURE 4-36 Half-wave rectification. (Courtesy Onan.)

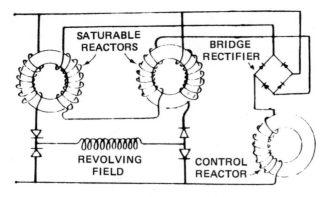

FIGURE 4-37   Amplifier circuit. (Courtesy Onan.)

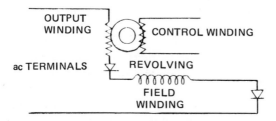

FIGURE 4-38   Half-wave control. (Courtesy Onan.)

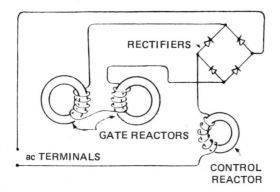

FULL WAVE CONTROL

FIGURE 4-39   Voltage regulator circuit. (Courtesy Onan.)

greater than desired. Therefore, it is necessary to add a control over the alternator field current to secure the rated output voltage. This is accomplished with the reactor control windings.

REACTOR   Control windings, when supplied with current of the correct polarity, act to demagnetize (or reset) the saturable reactor cores after they have been saturated by line current in the output windings. When demagnetized, the reactor cores oppose the line voltage. This takes place for some part of that one half-wave during which current is supplied to the alternator field. Thus part of the half-wave is supplying voltage to saturate the reactor core, and the other part sends current to the field coils. By adjusting the amount of current to the control windings, the amount of demagnetizing is controlled. This in turn precisely controls the amount of excitation field current and the resulting alternator output voltage. The adjustment of the amount of control current required is handled by the regulator.

REGULATOR The regulator consists of a saturating reactor (a toroid coil with only one winding), a rectifier bridge circuit, and connections to the control windings of the saturable reactors in the exciter circuit (Fig. 4-39). The saturating reactor connects to the rectifier bridge and the control windings of the saturable reactors.

SATURATING CONTROL REACTOR The saturating reactor is voltage sensitive and has the property of opposing line voltage up to a predetermined value. But when the line voltage exceeds that predetermined value, the reactor permits current to flow. If alternator output voltage is below the set voltage, no current passes through the coil. Therefore, no demagnetizing current flows to the control windings of the saturable reactors. This allows full current to reach the field coils, which in turn build up the alternator output voltage.

However, as the alternator output voltage comes up to requirement and then tends to exceed the set voltage, the saturating reactor in the regulator allows current to flow through its coil to the regulator bridge rectifiers.

The rectifier current then flows through the control windings of the saturable reactors. This in turn demagnetizes the saturable reactors. Demagnetizing the reactor cores results in opposing the flow of current to the alternator field coils and so reduces the alternator's output voltage back to the set requirements. Figure 4-40 shows the complete magneciter alternator circuit.

Regulation is literally instantaneous since rated voltage is restored within two seconds after being affected by a load change.

BRUSHLESS EXCITER AC GENERATOR The YD generators beginning with J-Series Spec. AA (Fig. 4-41) are four-pole, revolving field, brushless exciter, 4- or 12-lead, reconnectible models. Generator design includes both single- and three-phase 60- and 50-Hz-type generators (Fig. 4-42).

GENERATOR OPERATION The basic operation of the generator involves the stator, voltage regulator, exciter field and armature, a full-wave bridge rectifier, and the generator rotor. Residual magnetism

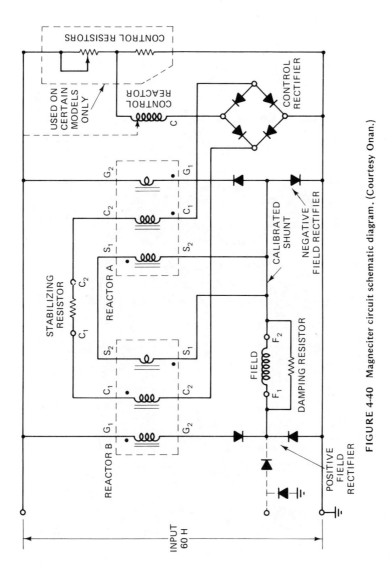

FIGURE 4-40 Magneciter circuit schematic diagram. (Courtesy Onan.)

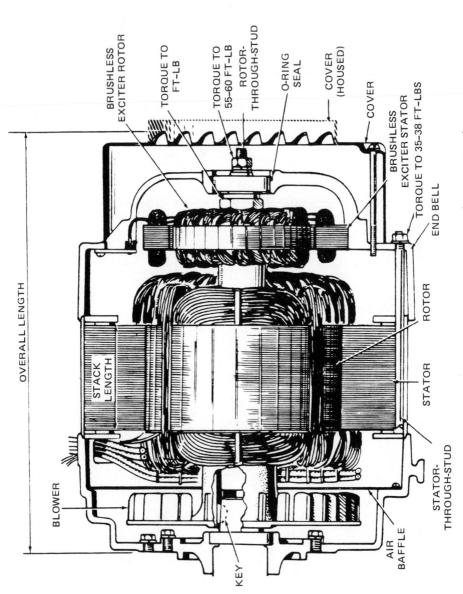

FIGURE 4-41 Sectional view of YD generator — J series. (Courtesy Onan.)

Labels:
- BRUSHLESS EXCITER ROTOR
- TORQUE TO FT-LB
- TORQUE TO 55-60 FT-LB ROTOR-THROUGH-STUD
- O-RING SEAL
- COVER (HOUSED)
- COVER
- BRUSHLESS EXCITER STATOR
- TORQUE TO 35-38 FT-LBS
- END BELL
- STATOR
- ROTOR
- STATOR-THROUGH-STUD
- AIR BAFFLE
- KEY
- BLOWER
- STACK LENGTH
- OVERALL LENGTH

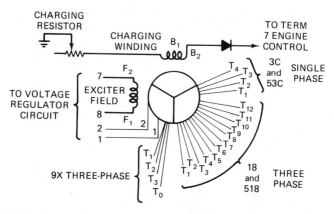

FIGURE 4-42 Single- and three-phase generator schematic (composite). (Courtesy Onan.)

in the generator rotor and a permanent magnet embedded in one exciter field pole begin the voltage buildup process as the generator set starts running. Single-phase ac voltage, taken from one of the stator windings, is fed to the voltage regulator as a reference voltage for maintaining the generator output voltage. The ac reference voltage is converted to dc by a silicon controlled rectifier bridge on the voltage regulator printed circuit board (Fig. 4-43) and fed into the exciter field windings. The exciter armature produces three-phase ac voltage that is converted to dc by the rotating rectifier assembly. The resultant dc voltage excites the generator rotor winding to produce the stator output voltage for the ac load.

The generator rotor also produces ac voltage in the charging winding of the stator which is converted to direct current for battery charging.

TYPES OF FUEL

There is a wide range of electric generator sets from 1 to 85 kW, gasoline and/or gaseous fuel. The advantages of these two types of fuel are discussed below. It should be remembered that each type of fuel is superior in some respects and less favorable in others. The following general comments should be considered in the light of individual preferences, local conditions, and fuel availability.

ADVANTAGES OF GAS OPERATION

MORE EFFICIENT OPERATION Because this fuel is already a gas, it mixes readily with the air, and combustion is more complete.

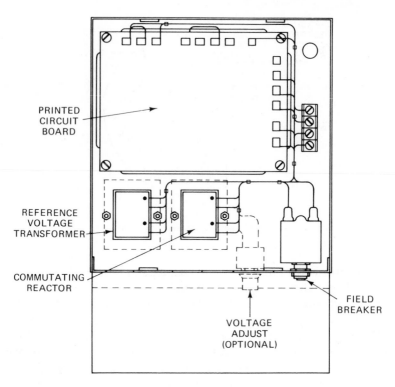

FIGURE 4-43 Voltage regulator assembly wiring diagram. (Courtesy Onan.)

LONGER LIFE   There is no lead in gas so deposits do not accumulate. The engine runs cleaner and lasts longer, requiring less maintenance and fewer oil changes.

QUICK STARTING Starting after long shutdowns is quicker with natural or LP gas because, unlike gasoline, gas remains "fresh" in storage.

BTU CONTENT IS IMPORTANT There are several different gaseous fuel mixtures: natural, manufactured, and bottle gas, or, as it is known commercially, butane or propane. Care should be taken to check the Btu content of gaseous fuels. The Btu content must be at least 1100 to get the same output as a similar gasoline set. Butane and propane gases meet this requirement. Some natural or manufactured gases, however, run as low as 450 Btu. A system using 450-Btu fuel must be derated 40 to 50%. For further information on gaseous fuel operation, write to the Onan factory for *Technical Bulletin No. T 015* or *Manual SP-1020*, 1400 73rd Ave. N.E., Minneapolis, Minn. 55432.

ADVANTAGES OF GASOLINE OPERATION

LOWER INITIAL COST A gasoline or gas set generally costs less than diesel.

QUICKER STARTING Gasoline and gas engines are usually easier to start under difficult conditions and in extremely low temperatures.

Diesel-driven electric generating sets are available in capacities ranging from 3 to 750 kW. Improved diesel engine design has made diesel fuel a reliable and dependable power source for most applications.

ADVANTAGES OF DIESEL OPERATION

COSTS LESS TO OPERATE Diesel fuel costs less per gallon and has a higher Btu content than gasoline. Fuel consumption is also much less as compared to gasoline. This efficiency and saving in fuel costs increase with the size of the unit.

REDUCED MAINTENANCE COSTS The heavier weight construction inherent in diesel design, plus the absence of points, plugs, and condensers, reduces service and repair needs by almost 50% over gasoline.

SAFE, LOW-VOLATILITY FUEL The high flash point and low volatility of diesel fuel reduces the possibility of fire or explosion from fumes and leakage, making underground storage unnecessary.

QUICK RESPONSE Diesels start promptly on diesel fuel alone without auxiliary devices, reach operating speed quickly, and handle lugging loads easily.

PLANNING AND SELECTION OF EQUIPMENT

When an emergency generating set is being considered, there is a tendency to skimp and cut corners on equipment on the premise that the system may never be used. Equipment that cannot handle the job is little better than none at all. When standby power is needed, it is desperately needed, and the best and most reliable equipment pays off at just such times.

CAPACITY The most important factor in selection of a standby system is, of course, size or capacity. A careful study should be made to

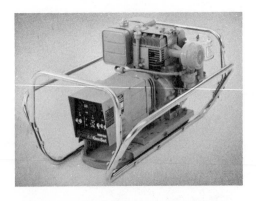

FIGURE 4-44 5.0 PKI portable gasoline. genset.

FIGURE 4-45 5.0 BGA RV genset.

FIGURE 4-46 6.5 MCCK Marine genset.

determine exactly what degree of service it is desirable to maintain during a power outage. How much and what kind of equipment must the system operate? What is the total wattage?

It is advisable at this stage to adopt a long-range view since it often becomes necessary to add more equipment to the load in future years. It may be desirable to provide full protection at the outset, since the

FIGURE 4-47   25.0 MOTA Marine
genset (diesel).

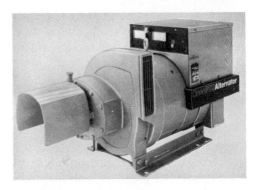

FIGURE 4-48   Power takeoff alternator.

FIGURE 4-49   750.0 DFZ diesel genset.

higher cost for a larger unit if offset by special circuit wiring costs necessary for selected protection. As needs grow, the set may still be made adequate by restricting protection to essential equipment. Of course, the decision of whether to provide full or selected protection will most often be decided by the particular operation for which the unit is provided. A hospital, for example, requires full protection for all essential equipment, while another building may only wish to maintain emergency lighting.

Another factor to consider at this stage is the altitude of your locale and the effect it may have on the rating of your standby set. If the point of installation is at a higher altitude than that of the manufacturer, the unit is generally derated 4%/1000 ft above sea level.

Also, complex factors, such as surge currents of electric motors, the magnetizing current of transformers, feedback from elevator loads, heating effects, and electromagnetic interference from solid-state loads, must be thoroughly explored and understood before system parameters are finalized.

FUEL Before the engine can be selected, careful consideration must be given to selecting the proper type of fuel. Often, the type of fuel is one of the most important considerations in final engine selection.

Fuel options are gasoline, LP gas, natural gas, and diesel. Specific advantages of each have been discussed earlier in this chapter. However, several general points which may outrule all others should be considered at the outset:

1. Availability of the various fuels in your particular location.

2. Local regulations governing the storage of gasoline and gaseous fuels.

COOLING Air and liquid cooling systems are available. Liquid cooling may be either a radiator or freshwater cooling system. The determining factor between air or water cooling will generally be the size of the unit. Air cooling must be confined to smaller units up through 15 kW. In these smaller sizes it is possible to force air over the engine and generator at a sufficient rate to cool it properly and thus take advantage of lower equipment and maintenance costs. Larger units, above 15 kW, should as a rule be equipped with a liquid system for maximum cooling protection.

Regardless of what capacity or cooling system your job requires, location, size, and temperature of the room must be considered. City water cooling, remote radiators, and heat exchangers are available for special applications.

LOAD TRANSFER SWITCH Electric utilities require transfer switches and usually supervise their installation, since power lines on which utility personnel work are affected.

There are two types of transfer switches: manual and automatic. The manual transfer switch must be operated by a person on duty either at the plant or a remote station. The plant is started by manual switch, and the load is transferred from one source to another by a hand-operated double throw switch.

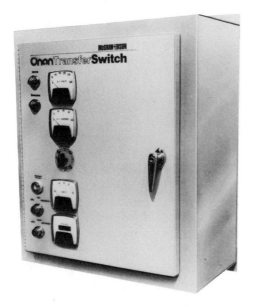

FIGURE 4-50 Onan OTII transfer switch.

FIGURE 4-51 Floor-standing OT transfer switch.

An automatic transfer switch starts the plant and transfers the load automatically, not requiring the attention of an operator. With automatic transfer switches the power outage can be limited to less than 10 sec. An automatic load transfer switch is almost a necessity for applications where uninterrupted power is of prime importance, such as a hospital, or where equipment is remotely located or where health or safety is at stake.

AUTOMATIC SEQUENTIAL PARALLELING

Automatic sequential paralleling of two or more electric generating sets performing as a single power source offers important advantages to the user, such as increased power capability, greater versatility in installation and operation, and, most important, the greater reliability of a multiple-unit system.

Illustrated in Fig. 4–50 is a typical automatic sequential paralleling emergency system for hospitals. This system, which conforms to NFPA-76A, consists of engine-generator sets, paralleling switchboards, a totalizing board, and automatic transfer switches featuring plug-in modular-type integrated circuit logic that monitors the total system and controls the operation of each individual generating set.

If utility power is interrupted, an automatic transfer switch initiates starting of all three sets (A, B, and C) simultaneously. When

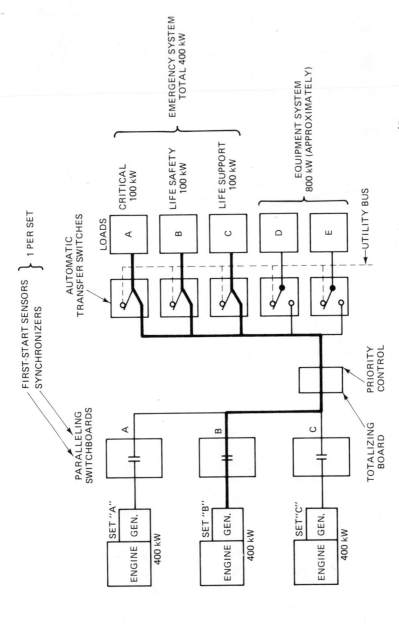

FIRST-START SENSORS
SYNCHRONIZERS } 1 PER SET

AUTOMATIC
TRANSFER SWITCHES

PARALLELING
SWITCHBOARDS

LOADS

CRITICAL
100 kW

LIFE SAFETY
100 kW

LIFE SUPPORT
100 kW

A

B

C

D

E

EMERGENCY SYSTEM
TOTAL 400 kW

EQUIPMENT SYSTEM
800 kW (APPROXIMATELY)

UTILITY BUS

A

B

C

PRIORITY
CONTROL

TOTALIZING
BOARD

SET "A"

SET "B"

SET "C"

ENGINE | GEN.

ENGINE | GEN.

ENGINE | GEN.

400 kW

400 kW

400 kW

FIGURE 4-52 Typical automatic sequential paralleling emergency system. (Courtesy Onan.)

74

the first of the three generating sets reaches operating voltage, a special emergency *bus* directs the power to the top-priority requirement areas, such as operating rooms and life safety and support systems—places where *life* may depend on fast power. The second and third sets are automatically synchronized to the emergency bus by a special circuit which permits their generating capacity to be directed to the less critical equipment load circuits.

With automatic sequential paralleling, each set generates power as soon as it starts, and any one can be the lead unit—or the first on the bus. Speed in assuming the major or critical load is a major advantage of this system. With all sets in the system answering a request for power, there is a much higher probability of at least one of them assuming the emergency load sooner than a single-set system.

Each generating set has its own synchronizer and electronically governed load sensor designed to automatically disconnect, shed, or add loads on a programmed priority basis. In the event of an engine-generator failure, a reverse power relay in that set senses the reverse power condition, opens its circuit breaker, and disconnects that load from the bus. Noncritical equipment loads are usually shed first, assuring continuous power to the critical life safety and support loads.

There are other advantages offered by sequential paralleling systems, such as automatic, unattended operation and system expansion. If an addition is built or power needs increase, the system can be expanded. When a component is serviced, it can be done with only a limited loss of power—rather than total power loss, as is the case with a single-unit source. Furthermore, a sequentially paralleled system can usually offer a cost advantage in dollars per kilowatts produced.

5

Conversion Equipment

Conversion substations and apparatus are those that change alternating current to direct current for certain specialized uses. Manual and automatic switching controls have been standardized for the main applications:

1. Electromechanical industries
2. Electric railways
3. Coal and ore mining
4. Special industrial dc power supplies

While much of this dc power is obtained directly from dc generators, synchronous converters, and batteries, the greatest part is derived from alternating-current sources through the use of rectifiers.

The application of rectifiers in industry ranges from very small amounts of rectified power, such as that needed for battery chargers and small dc motors, up to the hundreds of thousands of kilowatts needed for a large electrochemical process.

Other topics that will be covered in this chapter include equipment used to convert electrical systems from single phase to three phase.

RECTIFYING DEVICES

A rectifying device, in general, is an elementary device which has the property of effectively conducting current in only one direction. If a voltage of a certain polarity is applied to the rectifying device, the current will flow through the device in a certain direction which is called the forward direction. If a voltage of the opposite polarity is applied, the current will not flow through the rectifying device. Therefore, when an alternating current is applied to a rectifying device, the device will conduct current during one alternation of each cycle but will block, or stop, the current flow during the reversed alternation. The resulting current is an intermittent and pulsating current but one that is unidirectional, or rectified.

From the preceding paragraph, we can see how a rectifying device is used to obtain a direct current from an alternating-current source. To obtain the desired rectification, one single rectifier may be used, or a combination of the two—may be used.

All rectifying devices are designed to conduct current satisfactorily in one direction only. However, some types are more effective than others; that is, some types of rectifiers block reverse current completely, while others merely have lower ratios of forward to reverse conductivity.

TYPES OF RECTIFIERS

HIGH-VACUUM THERMIONIC RECTIFYING DEVICES   A hot cathode tube may be used as a rectifying device and consists of a highly evacuated tube with a heated cathode and an anode; it is sometimes known as a kenotron. The objective of this type of rectifier is to withstand very high inverse voltages, and such rectifiers are best adapted to applications requiring high dc voltages and small currents, such as X-ray applications and in testing certain types of insulation. Few are ever used anymore in the generation of electric power.

GASEOUS-DISCHARGE THERMIONIC RECTIFYING DEVICES Hot cathode tubes based on gaseous discharge have been filled with various gases, but mercury vapor seems to be the most popular. A small amount of mercury is introduced into the tube during manufacture and so located in the tube that it will be at a temperature near that of the ambient air. At this temperature, the mercury-vapor density is high enough to make plenty of mercury atoms available to keep the space charge down. At the same time, the mercury-vapor pressure is not high enough to break down readily when the anode voltage is negative. This type of tube has been used rather extensively for battery charging and is called a phanotron.

Another type of gaseous rectifier uses argon gas and a tungsten filament. This type of tube was once very popular for battery charging up to 100 V and 12 A.

A kenotron tube modified with the introduction of a wire-mesh screen or a metallic cylinder punched with holes and surrounding the cathode is called a pliotron. Such grid control is the basis for the operation of audio-frequency and radio-frequency amplifiers used in radio and TV in years past but has little use in the power generating field.

A gaseous-discharge tube designed with a grid is called a thyratron. Since the grid can be used to control the dc output voltage, thyratrons have been used quite extensively as a supply for small variable-speed motors and for other control functions. The usual operating voltage of thyratron tubes is under 600 V, and the current rating is 15 A or less per tube. Therefore, the practical horsepower limit for motor control is about 25 hp, but most are used on motors of less than 5 hp.

ARC-DISCHARGE RECTIFYING DEVICES   Rectifying devices utilizing the mercury pool as the cathode are known as mercury-pool tubes, and they operate on the arc-discharge principle. Most are enclosed in high-vacuum tubes or steel tanks.

SEMICONDUCTOR RECTIFYING DEVICES   One of the earliest semiconductors was the copper-oxide rectifying device, which is formed by oxidizing the surface of a copper plate. When voltage is applied between the oxidized surface and the copper, the current can be conducted readily from the oxidized surface to the copper but not in the other direction.

A later development was the selenium cell, or rectifying device. This device is formed by condensing selenium vapor on an aluminum plate in a vacuum and applying a conducting surface to the selenium. Selenium rectifying devices are built in the form of disks or plates up to about a foot square and around $\frac{1}{16}$ in. thick. Several devices may be mounted on a stud with insulating bushings and connected in series or parallel, or both, to form one or more circuit elements. Due to the large area of the element, such rectifiers are good at dissipating heat via their ventilating housing. In fact, with a moderate amount of forced ventilation, their current rating can be doubled.

Germanium rectifying devices are capable of increasing their current density over selenium devices more than 1000-fold—with an accompanying reduction in bulk. This device consists of a thin wafer of very pure germanium placed between thin layers of materials such as antimony or indium. The wafer is heated to a point where the indium and antimony diffuse into the surface of the germanium and is then enclosed in a hermetically sealed housing to exclude contaminants.

The silicon rectifying device is similar in size and physical appearance to the germanium rectifying device, but its allowable temperature limit is about twice that of the germanium device, making it more desirable for many applications.

RIPPLE

The output voltage of a rectifier contains a ripple component whose frequency depends on the number of phases of the rectifier. The magnitude of the ripple depends on the number of rectifier phases, the amount of phase control, and the loading of the rectifier. This ripple is of no consequence in applications where the rectifier has six phases or more. However, in some applications—such as in electric railway applications—it is sometimes necessary to filter the rectifier output and reduce the ripple to prevent interference with the communications system. In applications employing small motors with single-phase rectifiers, there may be overheating of the motors due to the ripple, and in such applications oversize motors are used.

RECTIFIER UNITS

An operative assembly of rectifiers used as a group—with the necessary rectifier auxiliaries, transformer equipment, and essential switchgear—is known as a rectifier group. In most installations there are one or more rectifier auxiliary devices, such as filament transformers, excitation equipment, cooling equipment vacuum pumps, and regulating equipment.

Many rectifiers require some means to maintain constant output voltage or current. This is often accomplished by means of phase control which is obtained by varying the level of a voltage or current in the excitation circuit. One way to accomplish this is to use variable resistance or a magnetic amplifier.

When semiconductors are connected in parallel, the problem of current division arises. This is caused by the volt-amp characteristics of individual cells differing from each other, making the impedances of the parallel circuit branches unequal.

When the rectifier voltage is so high that the PRV rating of a cell would be exceeded, it becomes necessary to connect two or more cells in series. In such cases, some cells will be subject to more voltage than others because their leakage currents are unequal. This condition may be corrected, however, by shunting each cell with a resistor.

MECHANICAL RECTIFIERS

Rectifiers falling under the mechanical category mainly include the synchronous motor, which is operated so that contacts are opened and closed at approximately the same rate that a rectifying element enters and leaves conduction with the same transformer connection. Voltage control can be achieved by shifting the phase of the driving motor, simulating grid control in a mercury-arc rectifier. The same result can also be achieved by suitable excitation of the commutation reactors. Best operation is obtained by avoiding the use of phase control and, instead, using some means to change the ac voltage, such as an induction regulator or a tap-changing transformer.

Since there is no rectifier element loss, mechanical rectifiers are often more desirable in industrial applications than other types. This is particularly important in the electrochemical field where the load is applied almost continuously.

RECTIFIER EQUIPMENT

Besides the rectifying elements, their circuits, and auxiliary equipment, a complete rectifier unit also includes transformers, saturable reactors, and overcurrent protection.

The transformer, of course, provides proper voltage to a rectifier when installed between the ac line and the rectifier. Transformers used in this way differ from distribution transformers in several respects.

1. The secondary kVA is greater than the primary kVA.

2. The reactance is usually greater than in a distribution transformer.

3. Stronger winding bracing is required in a single-way rectifier transformer than in other types.

Transformers for double-way circuits are very similar in winding arrangement to distribution transformers, except for the value of the secondary voltage. When semiconductor rectifiers are used, no surge protection is needed, nor is extra bracing.

Three general methods of rectifier voltage regulation are available: adjustment of the ac voltage supplied to the rectifier, adjustment of the impedance in the ac circuit, and phase control of the rectifier firing.

Nearly all electrical devices—including rectifier equipment—require protection against sustained overloading and ground faults. In addition, protection may be needed against specific dangers such as overspeed, loss of field, and incorrect phase sequence. In rectifier equipment the items most in need of overload protection—outside the rectifier auxiliaries—are the transformer in a single-way rectifier circuit and the cells in a semiconductor-rectifier circuit.

The overload and arc-back protection in the ac circuits of rectifier installations is in most instances obtained by circuit breakers. When the voltage is 600 V or less, air circuit breakers are used for protection against overloads. At higher voltages, high-voltage air circuit breakers or oil circuit breakers are used in the supply circuit. At extremely high voltages, the expense of suitable circuit breakers is sometimes avoided by using fuses and disconnecting switches in the supply.

PHASE CONVERTERS

It is sometimes desirable to use three-phase equipment on single-phase electric services. One such converter that will enable this connection is called the Roto-Phase manufactured by Arco Electric Products Corp. While called a phase converter, it is more accurately a phase generator because it generates one voltage which, when paralleled with the two voltages generated from a single-phase line, produces three-phase power. Induction as well as resistance three-phase loads can be operated from a single-phase supply.

A schematic diagram of a Roto-Phase connected to a single-phase source to operate two three-phase motors is shown in Fig. 5-1. In general, L_1 and L_2 are connected through a separate safety switch to Roto-Phase leads T_1 and T_2 at the junction box. The Roto-Phase lead, T_3, is then run through the third pole of the three-pole safety switch to the motor controls. L_1 and L_2, along with T_3, are then connected to

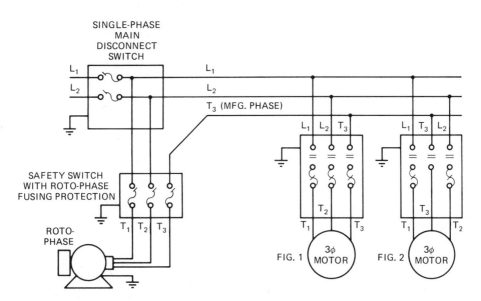

FIGURE 5-1 Schematic diagram of a Roto-Phase converter being used on a single-phase service to operate two three-phase induction motors. (Courtesy Arco Electric.)

the three-phase motor controls to provide three-phase service to the motors.

In using the Roto-Phase, do not connect any single-phase load or magnetic controls to T_3. This lead can be readily identified by the leg with the highest voltage to ground with the Roto-Phase running alone.

Should the motor starter have only two overload relays, do not run lead T_3 through a relay. Be sure to properly ground all electrical equipment according to the NE Code.

Always start the Roto-Phase before energizing motors and be sure to follow wire sizings carefully. Properly maintained voltages on motor starts are very important, so wire distances, sizes, and voltage drops should be studied carefully.

6

Primary Distribution Systems

The source of most commercial electric energy is a generator or a combination of generators. The generator is driven either by engines, hydraulics, or steam—although wind power is again being experimented with. Steam can be developed by coal (the most popular means at present), oil, gas, or nuclear fission.

Most generating stations utilize an outdoor substation, which contains step-up transformers to transform the generator voltage of, say, 14 kV to the transmission voltage, which may be as high as 750 kV. A three-phase transformer is normally used for each generator.

Other equipment contained in the outdoor substation includes disconnect switches for isolating circuit breakers, line-grounding disconnect switches, and lightning arresters. Current and potential instrument transformers (see Chapter 12) are also usually located in the outdoor substation to supply operating power for protective relays and metering devices.

The main transmission lines between the generating station and the bulk power station are usually protected by differential protective relays which measure the incoming and outgoing currents. The relays operate to open the line circuit breakers, usually placed at each end of the line, when a fault occurs and the incoming and outgoing currents do not balance. This way, any faults may be quickly isolated to main system stability and preserve service to the unfaulted portion of the system.

The outdoor bulk power substation, mentioned in the preceding paragraph, is very similar to the generating station substation in that it contains circuit breakers, the outdoor substation, and other related switchgear equipment. At the generating station substation, the generated voltage is stepped up to the transmission voltage. However, at the bulk power substation, this voltage must again be stepped down, or reduced, by the use of step-down transformers. From this point, power is supplied to distribution substations, and each of these subtransmission circuits are protected by some form of overcurrent protection.

Subtransmission and primary feeder circuits have many variations and voltages. The actual voltage used is governed by a large number of factors—mainly economic but also including such practical matters as the area to be served, load densities, estimated future growth, terrain, and availability of rights-of-way and substation sites.

Various designs of utilization substations are also needed to provide the many types of services that are in demand. Typical loads include residential services, which are usually 240/120 V, single phase, and commercial services, which may be 240/120 V, single phase, or perhaps 208/120 V, Y, three phase. Some larger installations are utilizing 480/208Y, three-phase services, while some industrial applications may require 13-kV service.

The local utility companies may require synchronous condenser or capacitor substations to maintain better system voltage, or some feeders may supply a low-voltage network where network protectors are used. Such circuits are usually equipped with some form of fault detector and with an automatic-reclosing switching scheme for automatic restoration of service after an outage.

Conversion substations are used in some areas to change alternating current to direct current for services such as 600, 1500, or 3000 V for railways such as used on the Pennsylvania Railway. Certain types of mills and electrochemical plants also require direct current for their manufacturing processes. Sometimes the plant itself provides this conversion equipment (see Chapter 5), while the local power companies provide the conversion in other cases.

PROTECTION OF POWER SYSTEM

A variety of situations may interfere with the normal operation of a power system. The predominant abnormal conditions on distribution circuits are line faults, system overloads, and equipment failures. Atmospheric disturbances and both animal and human interference with the system are generally the underlying causes of these conditions.

Line faults can be caused by strong winds which whip phase conductors together or which blow tree branches on the lines. In winter, freezing rain can produce a gradual buildup of ice on a circuit, and eventually one or more conductors may break and fall to the ground.

Squirrels and other animals have been known to place themselves between an energized portion of the circuit and ground, producing sad results for the animal as well as the circuit.

On underground systems, cables severed by earth-moving equipment are a prevalent cause of faults. Lightning strokes can fault a system by opening lines or initiating arcs between conductors. Unforeseen load growth is the primary cause of overloads. Equipment failure can be caused by lightning; insulation deterioration; improper design, manufacture, installation, or application; and system faults.

TYPES OF FAULTS

Line-to-ground, line-to-line, and double line-to-ground are faults common to single-, two-, and three-phase systems. The three-phase fault, naturally, is a characteristic only of the three-phase system.

Line-to-ground faults occur when one conductor falls to ground or contacts the neutral wire. Possible points along a distribution system where this type of fault can result are shown in Fig. 6-1.

Line-to-line faults can happen when conductors or a two-phase or three-phase system are short-circuited, as shown in Fig. 6-2. They can occur anywhere along a three-phase wye or delta system or along a two-phase branch.

Double line-to-ground faults occur when two conductors fall and are connected through ground or when two conductors contact the neutral of a three-phase or two-phase grounded system. Figure 6-3 shows a typical faulted circuit.

The various faults illustrated in Figs. 6-1 through 6-3 are all unsymmetrical. Faults such as these and other unbalanced conditions on polyphase systems are traditionally analyzed by the application of symmetrical component theory. This theory is adequately covered in a number of books on electrical theory, and due to the length required to cover it adequately, we shall omit it here.

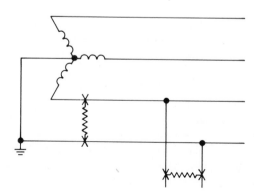

FIGURE 6-1 Line-to-ground faults.

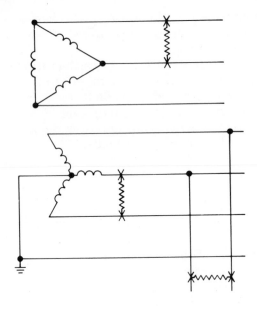

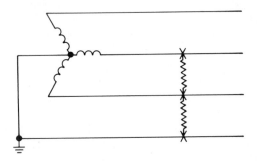

FIGURE 6-2   Line-to-line faults.

FIGURE 6-3   Double line-to-ground faults.

RELAY PROTECTION

Protective relays and their connections are many and varied. Adequate protection for a small distribution system connected to a few small generators may consist of time-cascaded overcurrent relays. This means that the most remote load may be protected by a straight solenoid-operated overcurrent relay that has no intentional time delay. When the current through the relay reaches overcurrent values, the relay operates to trip the breaker as fast as possible. The relays and breakers nearer the power source may operate on currents of fault magnitudes but with enough time delay to permit the more remote breaker to operate if it is going to. The relays still nearer the source of generation will operate on the same principle but with longer and longer time delays. Time separations of one-quarter of a second are adequate in most cases.

Large power systems fed from large generating units require a different type of protecting system. Large systems denote large inertias which cannot be changed rapidly without seriously endangering the stability of the system. High-speed relays that can quickly detect the location of the fault and breakers that can open the fault current in about one-twentieth of a second are now a standard combination.

TRANSFER SCHEMES

Recent developments in distribution protective equipment, primarily the advent of electronically controlled recloser and sectionalizer controls, have allowed the usage of more complex, automated loop-type distribution systems. This development, along with more advanced load transfer schemes and supervisory equipment, has led to substantial improvements in the reliability of modern distribution systems.

SWITCHED LOAD TRANSFER SCHEMES A line schematic of a simple load transfer scheme using a McGraw-Edison Type S transfer control and electrically operated oil switches is shown in Fig. 6-4.

In this load transfer scheme, power is normally supplied from a preferred source and automatically switched to an alternate, standby source if the preferred source is lost for any reason. Upon restoration of preferred source voltage, the load is switched back automatically or manually. Return switching can be either in a closed-transition (parallel return) mode (the preferred source closes before the alternate source opens) or in an open-transition mode (nonparallel return; the alternate source opens before the preferred source closes).

The type S control is designed for use with three single-phase switches or one three-phase switch in each line; either single-phase or three-phase sensing can be employed.

LOAD TRANSFER—MANUAL RETURN This scheme (Fig. 6-5) uses electronically controlled reclosers in both the preferred source and alternate source lines. In these examples, reclosers are equipped with McGraw-Edison Type LS controls. Darker lines on illustrations represent energized lines. Load is normally fed from the preferred source, S1.

Recloser ACR1 is normally closed and senses voltage (one or three phases) on its source side. It opens after a time delay upon loss of S1 voltage.

Recloser ACR2 is normally open and senses voltage (one or three phases) on its load side. It closes after a time delay (longer than ACR1) upon loss of load voltage. In addition, ACR2 is equipped with a block-of-reclose accessory energized from the alternate source (S2) which prevents any attempt to close ACR2 if S2 is not energized.

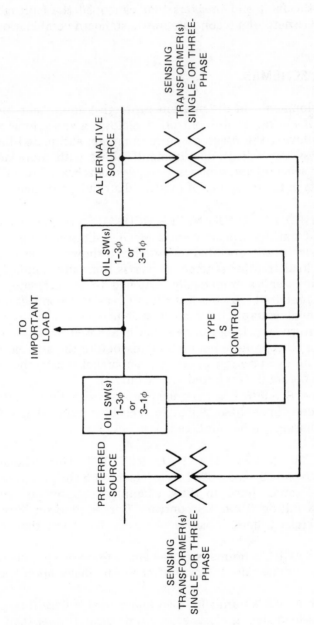

FIGURE 6-4 Line schematic of a simple load transfer scheme. (Courtesy McGraw-Edison.)

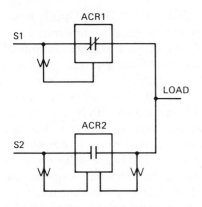

FIGURE 6-5 (Courtesy McGraw-Edison.)

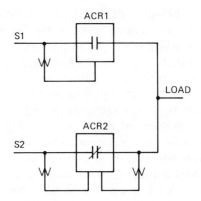

FIGURE 6-6 (Courtesy McGraw-Edison.)

When preferred source voltage (S1) is lost, the controls of both reclosers sense the loss of voltage. If voltage is not restored within the time delay selected, ACR1 opens.

After a longer time delay, ACR2 closes to restore service to the load (Fig. 6-6). A momentary cold-load pickup accessory can be provided for ACR2 to prevent tripping on cold-load inrush.

When preferred source voltage (S1) is restored, transfer back to the preferred source is accomplished manually.

If a permanent fault occurs on the load side of the system as shown in Fig. 6-7, the preferred source recloser (ACR1) operates to lockout. The alternate source recloser (ACR2) senses the loss of load voltage, and after a time delay, ACR2 closes into the fault and also operates to lockout, as shown in Fig. 6-8. (A momentary nonreclose accessory is available for one-shot lockout of ACR2 to minimize load disturbances.) After the fault is cleared, service from the preferred source is restored manually.

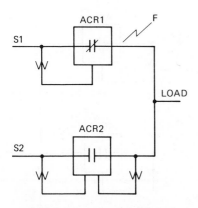

FIGURE 6-7 (Courtesy McGraw-Edison.)

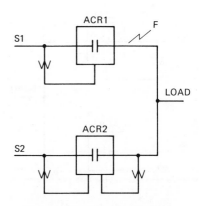

FIGURE 6-8 (Courtesy McGraw-Edison.)

LOAD TRANSFER—AUTOMATIC RETURN Upon loss of preferred source voltage the load is automatically transferred to an alternate source. When preferred source voltage is restored, the load is automatically transferred back to the preferred source. The scheme shown in Fig. 6-9 uses electronically controlled reclosers in both the preferred source and alternate source lines. Both reclosers are equipped with type LS controls or with a McGraw-Edison Type S control. A requirement of this scheme is that the reclosers must be near enough to establish a communication link between them. Load is normally fed from the preferred source, S1. Recloser ACR1 is normally closed and senses voltage (one or three phases) on its source side. It opens after a time delay upon loss of S1 voltage. The control of the normally open alternate source recloser (ACR2) is connected to the control of ACR1. In addition, ACR2 is equipped with a block-of-reclose accessory energized from the alternate source (S2) which prevents any attempt to close ACR2 if S2 is not energized.

When preferred source voltage (S1) is lost, the control of ACR1 senses the loss of voltage. If voltage is not restored within the time delay selected, ACR1 opens and signals ACR2 to close to restore service to the load, provided the alternate source (S2) is energized, as shown in Fig. 6-10. (A momentary cold-load pickup accessory can be provided for the alternate source recloser to prevent tripping on cold-load inrush).

When preferred source voltage (S1) is restored, return to preferred source is automatic through either parallel return (ACR1 closes before ACR2 opens), as shown in Fig. 6-11, or nonparallel return (ACR2 opens before ACR1 closes), as shown in Fig. 6-12. The system is restored to normal.

If a permanent fault occurs on the load side (Fig. 6-13), the preferred source recloser (ACR1) operates to lockout. However, the alternate source recloser (ACR2) is blocked from closing due to the presence of preferred source voltage (S1) and/or the lockout state of ACR1.

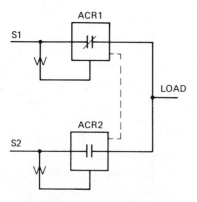

FIGURE 6-9 (Courtesy McGraw-Edison.)

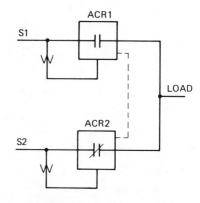

FIGURE 6-10 (Courtesy McGraw-Edison.)

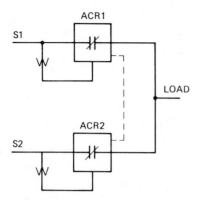

FIGURE 6-11 (Courtesy McGraw-Edison.)

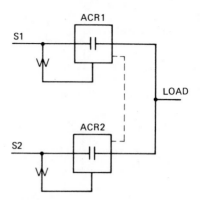

FIGURE 6-12 (Courtesy McGraw-Edison.)

After the fault is cleared, the transfer scheme is reset by reclosing ACR1.

LOOP SCHEMES   Loop schemes have been developed to improve reliability and maintain service continuity to the greatest number of customers. In a loop scheme, two distribution circuits are tied together by a normally open recloser equipped with a tie control. To further improve continuity, a normally closed recloser equipped with a sectionalizing control is added near the midpoint of each line. In these examples, the McGraw-Edison Type LS control accessories are used.

LOOP SCHEME WITH THREE RECLOSERS   In the simplest loop scheme, three electronically controlled reclosers are employed, as shown in Fig. 6-14. ACR1 and ACR2 are normally closed reclosers

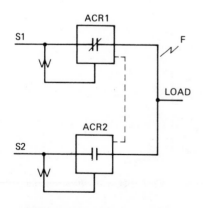

FIGURE 6-13 (Courtesy McGraw-Edison.)

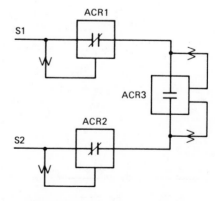

FIGURE 6-14 (Courtesy McGraw-Edison.)

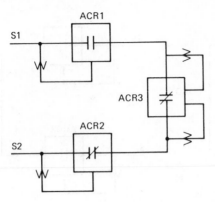

FIGURE 6-15 (Courtesy McGraw-Edison.)

equipped with type LS sectionalizing controls; they open upon loss of their respective source voltage after a time delay.

ACR3 is a normally open recloser equipped with a type LS tie control; it closes upon loss of voltage on either side after a time delay (longer than ACR1/ACR2). Upon loss of S1 voltage, both ACR1 and ACR3 sense the loss of voltage. If voltage is not restored within the time delay selected, recloser ACR1 opens. After a longer time delay, recloser ACR3 closes. The entire loop (up to ACR1) is then fed from source S2 (Fig. 6-15). Return to normal is manual.

If a permanent fault occurs on the load side of ACR1 (Fig. 6-16), ACR1 operates to lockout. The tie recloser ACR3 senses the loss of voltage and after a time delay closes into the fault and also operates to lockout. The faulted section is isolated, and service is maintained to the balance of the loop, as shown in Fig. 6-17. (A momentary nonreclosing accessory can be added to ACR3 to provide one shot to lockout in the event of closing into a fault.)

LOOP-SECTIONALIZING SCHEME WITH FIVE RECLOSERS In this scheme (Fig. 6-18) each distribution circuit is divided into two

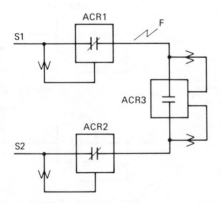

FIGURE 6-16 (Courtesy McGraw-Edison.)

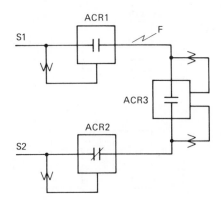

FIGURE 6-17 (Courtesy McGraw-Edison.)

sections of equal load through normally closed feeder reclosers. Each circuit is connected at the tie point with a normally open tie recloser. The reclosers are set to isolate a section selectively under permanent fault and to transfer the unfaulted sections to the adjacent circuit.

ACR1 and ACR2 are normally closed reclosers equipped with McGraw-Edison Type LS sectionalizing controls. They open upon loss of source voltage after a time delay.

ACR3 and ACR4 are also normally closed reclosers equipped with type LS sectionalizing controls. However, upon loss of source voltage and after a time delay (longer than ACR1/ACR2) they change their minimum trip value and number of shots to lockout.

ACR5 is a normally open recloser equipped with a type LS tie control. It closes upon loss of voltage on either side after a time delay (longer than ACR3/ACR4).

Upon loss of S1 voltage, reclosers ACR1, ACR3, and ACR5 sense the loss of voltage. If voltage is not restored within the time delay selected, ACR1 opens. After an additional time delay, ACR2 changes its

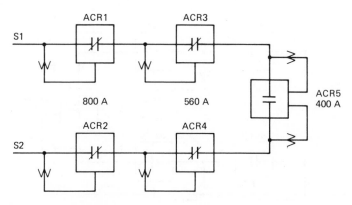

FIGURE 6-18 (Courtesy McGraw-Edison.)

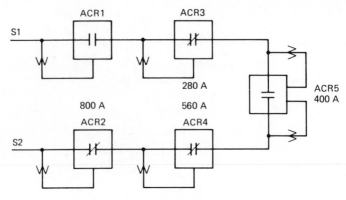

FIGURE 6-19 (Courtesy McGraw-Edison.)

minimum trip value from 560 to 280 A and changes to one shot to lockout to coordinate with the tie recloser, ACR5, as shown in Fig 6-19.

After a longer time delay than ACR3, recloser ACR5 closes; the entire loop up to ACR1 is fed from source S2, as shown in Fig. 6-20. Return to normal is accomplished manually.

If a permanent fault occurs at F1 (Fig. 6-21), recloser ACR1 operates to lockout. Reclosers ACR3 and ACR5 sense the loss of voltage. ACR3 times out and changes its minimum trip from 560 to 280 A and changes to one shot to lockout. After its time delay (longer than ACR3), ACR5 closes into the fault, and ACR3 locks out (Fig. 6-22). The fault is isolated; service is maintained to three-fourths of the loop.

If a permanent fault occurs at F2 (Fig. 6-21), recloser ACR3 operates to lockout. Recloser ACR5 senses the loss of voltage and, after its time delay, closes into the fault and operates to lockout (Fig. 6-23). The fault is isolated and service is maintained to three-fourths of the loop.

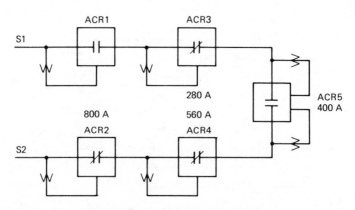

FIGURE 6-20 (Courtesy McGraw-Edison.)

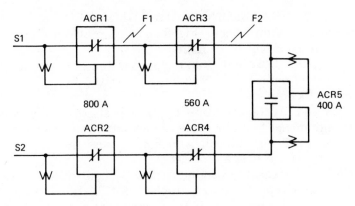

FIGURE 6-21 (Courtesy McGraw-Edison.)

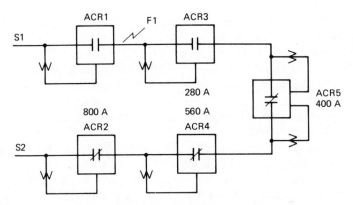

FIGURE 6-22 (Courtesy McGraw-Edison.)

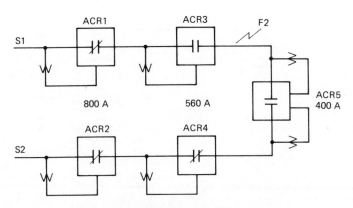

FIGURE 6-23 (Courtesy McGraw-Edison.)

LOOP-SECTIONALIZING SCHEME WITH COUNTING TIE This is a three-recloser loop scheme (Fig. 6-24) with provision to prevent the tie recloser from closing into a fault.

ACR1 and ACR2 are backup reclosers or breakers and not directly involved in the loop scheme. They must, however, be set for less than four shots to lockout.

ACR3 and ACR4 are normally closed, electronically controlled reclosers and are set for four shots to lockout. They are equipped with type LS sectionalizing controls, which will cause them to open upon loss of source voltage after a time delay.

ACR5 is a normally open, electronically controlled recloser. It is equipped with a type LS tie control which will also count each time voltage is lost on either side. If four counts are registered in a preset time interval, closing of ACR5 is blocked. If, however, three or fewer counts are registered within the time interval, the control will close ACR5.

Operation of the scheme for the loss of source voltage is the same as for a basic three-recloser loop scheme.

If a permanent fault occurs at F1 (Fig. 6-24), ACR1 operates its three shots to lockout. Reclosers ACR3 and ACR5 sense the loss of voltage, and ACR5 registers three counts. After its time delay ACR3 opens, and after an additional time delay, tie recloser ACR5 closes since it counted only three times. The fault is isolated, and service is maintained to three-fourths of the loop (Fig. 6-25).

If a permanent fault occurs at F2 (Fig. 6-24), ACR3 operates its four shots to lockout. ACR5 senses the loss of voltage during operation of ACR3, registers four counts, and is blocked from closing into the fault. The fault is isolated; service is maintained to three-fourths of the loop, and the unfaulted portion of the circuit experiences a minimum of disturbances.

If the fault at F2 is a three-phase fault, recloser ACR3 operates its

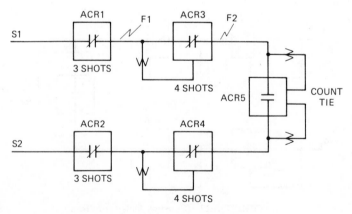

FIGURE 6-24 (Courtesy McGraw-Edison.)

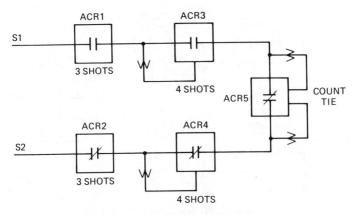

FIGURE 6-25 (Courtesy McGraw-Edison.)

four shots to lockout. However, the voltage does not recover each time the recloser operates so the control of the tie recloser (ACR5) registers only one count, and after a time delay ACR5 closes into the fault. ACR5 operates to lockout. The fault is isolated, and service is maintained to three-fourths of the loop (Fig. 6-26).

LOOP-SECTIONALIZING SCHEME WITH ELECTRONIC SECTIONALIZERS AT THE MIDPOINT OF EACH LINE   By substituting electronically controlled sectionalizers for reclosers at the midpoint of each line, the following loop scheme is developed. This arrangement would probably be limited to those cases where another step in coordination is not possible. See Fig. 6-27.

ACR1 and ACR2 are normally closed, electronically controlled reclosers equipped with type LS sectionalizing controls. They open upon loss of voltage after a time delay.

SEC1 and SEC2 are normally closed, electronically controlled

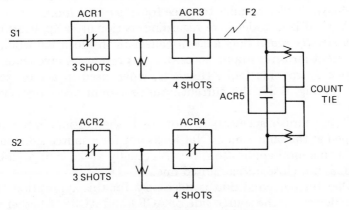

FIGURE 6-26 (Courtesy McGraw-Edison.)

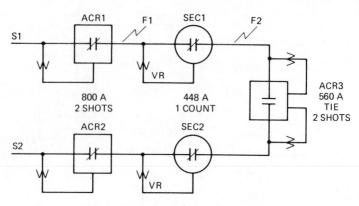

FIGURE 6-27 (Courtesy McGraw-Edison.)

sectionalizers, operating as normal sectionalizers. That is, they are equipped with voltage restraint only and do not have type LS controls. In this example, they are set for one count. Their minimum actuating current is selected to coordinate with tie recloser ACR3.

ACR3 is the normally open, electronically controlled tie recloser equipped with a type LS tie control. It will close upon loss of voltage on either side after a time delay (longer than ACR1/ACR2).

Upon loss of S1 voltage, both ACR1 and ACR3 sense the loss of voltage. If voltage is not restored within the time delay selected, ACR1 opens. After an additional time delay, tie recloser ACR3 closes, and the loop is fed from source S2.

If a permanent fault occurs at F1 (Fig. 6-27), ACR1 operates twice and locks out. Tie recloser ACR3 senses the loss of voltage, and after a time delay, ACR3 closes into the fault.

During the first trip operation of recloser ACR3, the downline sectionalizer SEC1 counts and opens. Sectionalizer SEC2 is blocked from counting by its voltage restraint accessory (source voltage is not lost). The fault is isolated, and ACR3 recloses into an unfaulted line (Fig. 6-28). Service remains to three-fourths of the loop.

A word is in order here regarding sectionalizer application. If this were a *strictly radial* line, a sectionalizer in this location would require the current-inrush-restraint accessory to prevent an erroneous opening due to a source-side fault. However, when used in a loop scheme as shown in Fig. 6-27, the current-inrush-restraint accessory should *not* be used.

If the inrush accessory is not used, the sectionalizer may count and open at the same time ACR1 locks out for a source-side fault. However, in this loop application, this is actually an advantage since ACR3 then does not close into a faulted line.

One further condition is necessary for this application. The protective device on the source side of ACR1 and ACR2, whether a station

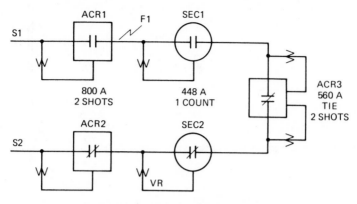

FIGURE 6-28 (Courtesy McGraw-Edison.)

breaker, station recloser, or another line recloser, must be set for the
same, or fewer, number of shots to lockout as the count setting of the
sectionalizer to prevent erroneous counting and opening of the sectionalizer for a fault on the source side of reclosers ACR1 and ACR2.

If a permanent fault occurs at F2 (Fig. 6-27), both SEC1 and
ACR1 sense the overcurrent. Recloser ACR1 operates, and during its
first trip operation, sectionalizer SEC1 counts and opens. ACR1 closes
into an unfaulted line. Tie recloser ACR3 senses the loss of voltage on
its S1 side and after a time delay closes into the fault.

ACR3 operates to lockout to isolate the fault and maintain service to three-fourths of the loop (Fig. 6-29). Sectionalizer SEC2 senses
the fault current during the operation of ACR3, but it is blocked from
counting by its voltage restraint accessory.

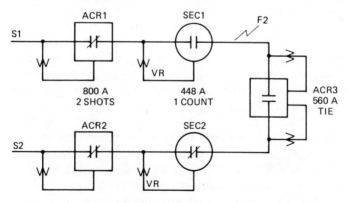

FIGURE 6-29 (Courtesy McGraw-Edison.)

7

Secondary Distribution Systems

Secondary distribution systems are installed and protected in a similar fashion to the higher-voltage primary distribution systems. The more important lines in the secondary system will require differential protection, while the less important ones may need nothing more than conventional overcurrent protection. The same rules apply, however, in that a fault must be quickly removed from the system by tripping the minimum number of circuit breakers or blowing the minimum number of fuses, leaving the balance of the system in operation.

There are two general arrangements of transformers and secondaries used. The first arrangement is the sectional form, in which a unit of load, such as one city street or city block, is served by a fixed length of secondary, with the transformer located in the middle. The second arrangement is the continuous form where the secondary is installed in one long continuous run—with transformers spaced along it at the most suitable points. As the load grows or shifts, the transformers spaced along it can be moved or rearranged, if desired. In sectional arrangement, such a load can be cared for only by changing to a larger size of transformer or installing an additional unit in the same section.

One of the greatest advantages of the secondary bank is that the starting currents of motors are divided between transformers, reducing voltage drops and also diminishing the resulting lamp flicker at the various outlets.

In the sectional arrangement, each transformer feeds a section of the secondary and is separate from any other. If a transformer becomes overloaded, it is not helped by adjacent transformers; rather, each transformer acts as a unit by itself. Therefore, if a transformer fails, there is an interruption in the distribution service of the section of the secondary distribution system that it feeds. This is the most often used layout for secondary distribution systems at the present time.

Power companies all over the United States are now trying to incorporate networks into their secondary power systems, especially in areas where a high degree of service reliability is necessary. Around cities and industrial applications, most secondary circuits are three phase—either 120/208Y or 480/208 V, Y connected. Usually, two to four primary feeders are run into the area, and transformers are connected alternately to them. The feeders are interconnected in a grid, or network, so that should any feeder go out of service, the load is still carried by the remaining feeders.

To protect a grid-type power system, a network protector is usually installed between the transformer and the secondary mains. This protector consists of a low-voltage circuit breaker controlled by relays, which cause it to open when reverse current flows from the secondaries into the transformer and to close again when normal conditions are restored. If a short circuit or ground fault should occur on a primary feeder or on any transformer connected to it, the feedback of current from the network into the fault through all the transformers on that feeder will cause the protective switches to open, disconnecting all ties between that feeder and the secondary mains. When the trouble is repaired and normal voltage conditions are restored on that feeder, the switches will reclose, putting the transformers back into service.

Designers of a network are always cautious about the placement of transformers. The transformers should be large enough and close enough together to be able to burn off a ground fault on the cable at any point. If not, such a fault might continue to burn for a long time.

The primary feeders supplying networks are run from substations at the usual primary voltage for the system, such as 4160, 4800, 6900, or 13,200 V. Higher voltages are practicable if the loads are large enough to warrant them.

Network power systems are usually installed underground with primaries, secondaries, and transformers all underground. The transformers and secondaries may, however, be overhead, or they may use a combination of overhead and underground construction.

SECONDARY SERVICES

The 4160-V transformer came about through the 4160Y-V connection on 2400-V transformers. In some cases, it was advantageous to connect transformers between phase wires on a 2400/4160Y-V system, and this required a transformer having a winding voltage of 4160 V. These 4160-

V transformers are now used in several ways. First, they are used in three-phase delta banks connected to 2400/4160Y-V systems.

Another application is on 4160-V single-phase lines taken off of a 2400/4160Y-V three-phase system, necessitating the use of 4160-V transformers.

In some instances, the 4160-V transformers are used for rural systems rated 4160/7200Y. With this system, 4160-V transformers can be used between phase wire and neutral of a three-phase four-wire system, and 7200-V transformers can be used between phase wires.

The 4800-V transformers are frequently used in some sections of the United States where distribution circuits run through thickly populated rural and suburban areas. Distribution lines in these localities are necessarily much longer than in cities, and therefore the 2400-V system is not high enough voltage to be economical. On the other hand, 4800-V distribution systems in these areas have proved to be quite logical and satisfactory.

Again, the systems originally were 4800-V delta, three-phase systems with 4800-V, single-phase branch lines. These delta systems, however, are now being converted, in many cases, to 4800/8320Y, giving a higher system voltage but using the same equipment that was used on 4800-V delta systems.

Rural electrification in thinly populated areas required still higher voltage for good performance and economy. Therefore, for rural power systems in certain sections of the United States, 7200-V distribution systems have been used quite extensively and successfully. The early rural systems were 7200 V delta, three phase in most cases, with 7200-V branch lines. These systems are now giving way to 7200/12,470Y-V, three-phase, four-wire systems. In fact, this system is probably the most popular in use today.

Although less popular than the 7200-V class of transformers, 7620-V systems are sometimes used for rural electrification. Most of these systems are actually 7620/13,200Y, three-phase, four-wire systems. On this type of system, 7620-V single-phase transformers can be used between phase wire and neutral of the three-phase system, or 13,200-V transformers can be used between phase wires. This type of system works out very economically for power companies that have both 7620- and 13,200-V distribution systems. In this situation, transformers can be used on either system, thereby making stocks of transformers flexible.

There are some 12,000-V, three-phase delta systems that were installed some time ago for transmission and power over greater distances than were feasible in lower voltages. There are now two applications for 12,000-V transformers. The first is for use on 12,000-V delta systems and the second for use on 7200/12,470Y-V systems.

The 13,200-V transformers also have two applications. First, they can be used on distribution systems that are 13,200 V delta, three phase, which were built to distribute electrical energy at considerable distance. The second application has already been mentioned in connection with the 7620/13,200-V, three-phase, four-wire system. On this system, the 13,200 standard transformer can be used between phase wires of the three-phase four-wire system. This connection is made quite often when it is necessary to connect a three-phase bank of transformers to the 7620/13,200Y-V systems.

In addition to use in rural areas, the 12,000- and the 13,200-V distribution systems are quite often used in urban areas. In relatively large cities having considerable industrial loads, 13,200- and 12,000-V lines are quite often run to serve industrial loads, while the 2400/4160Y-V system is used for the residential and commercial loads.

24,940-Grd-Y/14,400-V units have one end of the high-voltage winding grounded to the tank wall and are suitable only for use on systems having the neutral grounded throughout its length.

System voltages to 68 kV have been designated as distribution, although transmission lines also operate at these same voltages. The trend is to convert these lines to four-wire distribution systems and use transformers with primary windings connected in wye. As an example, a multigrounded neutral is added to a 34,500-V system, and 20,000-V transformers are used to supply the customers. The system is also in use for new construction in high-load-density areas.

USE OF CAPACITORS

A capacitor is a device that will accept an electrical charge, store it, and again release the charge when desired. In its simplest form, it consists of two metallic plates separated by an air gap. The larger the surface of the plates and the closer they are together, the greater the capacity will be.

If the space between the plates is filled with various insulating materials, such as kraft paper, linen paper, or oil, the capacity will be greater than with air. The increase in this capacity for any specific material as compared to air is called the dielectric constant and is expressed by the letter K. If plate area and spacing remain unchanged but a certain grade of paper gives a capacitor twice the capacity it had with air, then the value for K would be 2. All insulating materials have a value of K which merely expresses its dielectric effectiveness as compared to air. Thus, only three factors govern capacity:

1. How big are the plates?
2. How close are they?
3. What material separates them?

Capacitors operate on both direct- and alternating-current circuits. If connected to the terminals of a dc circuit, a pair of plates without a charge on them will accept a static charge. Current will rush into the capacitor until each plate is at the potential of the line to which it is connected. Once this potential is reached (a very rapid process), no further current, other than leakage current, will flow. If removed from the line, the capacitor plates will maintain their charge until it is dissipated by leakage between plates or by deliberate contact between plate terminals. This principle is used in surge generators. In the ac application, the plates are alternately charged and discharged by the voltage changes of the circuit to which they are connected. It is this condition which makes possible the use of capacitors for power-factor correction.

POWER FACTOR

Power factor is a ratio of useful working current to total current in the line. As power is the product of current and voltage, the power factor can also be described as a ratio of real power to apparent power and be expressed as

$$\text{Power factor} = \frac{\text{kW}}{\text{kVA}}$$

Apparent power is made up of two components, namely, real power (expressed in kilowatts) and the reactive component (expressed in kilovars or kvar). This relationship is shown in Fig. 7-1. The horizontal line AB represents the useful real power (or kilowatts) in the circuit. The line BC represents the reactive component (kvar) as drawn in a downward direction. Then a line from A to C represents apparent power (kilovolt-amperes). To the uninitiated, the use of lines representing quantity and direction is often confusing. Lines so utilized are called *vectors*. Imagine yourself at point A desiring to reach point C, but, due to obstructions, you must first walk to B, and then turn a right angle and walk to C. The energy you dissipated in reaching C was increased because you could not take the direct course, but, in the final

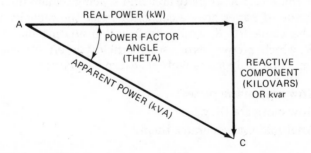

FIGURE 7-1 Power-factor triangle used to show angle between real power and apparent power.

analysis, you ended up at a point the direction and distance of which can be represented by the straight line AC. The power-factor angle shown is called theta.

There is interest in power factor because of the peculiarity of certain ac electrical equipment requiring power lines to carry more current than is actually needed to do a specific job unless we utilize a principle that has long been understood but not until recent years given the attention it deserves. This principle utilizes the application of capacitors.

LEAD AND LAG

We cannot add real and reactive components arithmetically. To understand why, consider the characteristics of electrical circuits. In a pure resistance circuit, the alternating voltage and current curves have the same shape, and the changes occur in perfect step, or phase, with each other. Both are at zero with maximum positive peaks and maximum negative peaks at identical instants. Compare this with a circuit having magnetic characteristics involving units such as induction motors, transformers, fluorescent lights, and welding machines. It is typical that the current needed to establish a magnetic field lags the voltage by $90°$.

Visualize a tube of toothpaste. Pressure must be exerted on the tube before the contents oozes out. In other words, there is no flow until pressure is exerted, or, analogously, the flow (current) lags the pressure (voltage). In like manner, the magnetic part of a circuit resists, or opposes, the flow of current through it. In a magnetic circuit, the pressure precedes or leads the current flow, or, conversely, the current lags the voltage.

Peculiarly, in a capacitor, we have the exact opposite: Current leads the voltage by $90°$. Visualize an empty tank to which a high-pressure air line is attached by means of a valve. At the instant the valve is opened, a tremendous rush of air enters the tank, gradually reducing in rate of flow as the tank pressure approaches the air line pressure. When the tank is up to full pressure, no further flow exists. Accordingly, you must first have a flow of air into the tank before it develops an internal pressure. Consider the tank to be a capacitor and the air line to be the electrical system. In like fashion, current rushes into the capacitor before it builds up a voltage, or, in a sense, the current leads the voltage.

A water system provides a better comparison. The system shown in Fig. 7-2 is connected to the inlet and outlet of a pump. For every gallon that enters the upper section, a gallon must flow from the lower section as the plunger is forced down. If the pressure on the upper half is removed, the stored energy in the lower spring will return the plunger to midposition. When the direction of flow is reversed, the plunger travels upward, and if flowmeters were connected to the inlet and out-

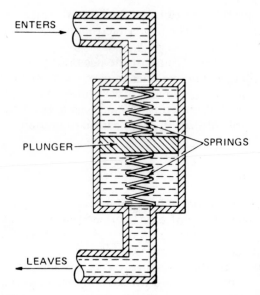

ENTERS

PLUNGER

SPRINGS

LEAVES

FIGURE 7-2 Water system used to show comparison of real power to apparent power.

let, the inlet and outlet flows would prove to be equal. Actually, there is no flow through the cylinder but merely a displacement. The only way we could get a flow through the cylinder would be by leakage around the plunger or if excessive pressure punctured a hole in the plunger.

In a capacitor, we have a similar set of conditions. The electrical insulation can be visualized as the plunger. As we apply a higher voltage to one capacitor plate than to its companion plate, current will rush in. If we remove the pressure and provide an external conducting path, the stored energy will flow to the other plate, discharging the capacitor and bringing it to a balanced condition in much the same fashion as the spring returned the plunger to midposition.

Since no insulator is perfect, some leakage current will flow through it when a voltage differential exists. If we raise the voltage to a point where we break down or puncture the insulation, then the capacitor is damaged beyond repair and must be replaced.

Cognizance of the three different kinds of current discussed, namely, in phase, lagging, and leading, permits the drawing of the relationship in Fig. 7-3.

The industry accepts a counterclockwise rotation about point M as a means of determining the relative phase position of voltage and current vectors. These may be considered as hands of a clock running in reverse. MC is preceding MR; hence, it is considered leading. MI follows MR; therefore, it is lagging.

The two angles shown are right angles ($90°$); therefore, it becomes apparent that MC and MI are exactly opposite in direction and will cancel out each other if of equal value.

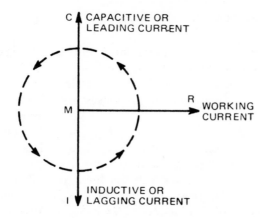

FIGURE 7-3 Phasor diagram used to compare leading current, working current, and lagging current.

Consider a wagon to which three horses are hitched, as in Fig. 7-4. If C and I pull with equal force, they merely cancel one another's effort, and the wagon will proceed in the direction of R, the working horse. If only R and I are hitched, the course will lie between these two.

Most utility lines contain quantities of working current and lagging current, and one type is just as effective in loading up the line as the other. If we know how much inductive current a line is carrying, then we can connect enough capacitors to that line to cancel out this wasteful and undesired component.

Just as a wattmeter will register the kilowatts in a line, a varmeter will register the kvar of reactive power in the line. If an inductive circuit is checked by a meter which reads 150,000 var (150 kvar), then application of 150-kvar capacitor would completely cancel out the inductive component, leaving only working current in the line.

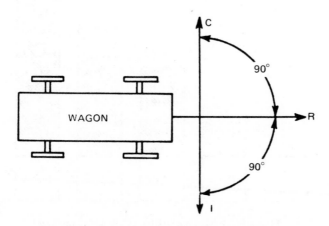

FIGURE 7-4 Horse and wagon comparison. See the text.

GRAPHIC RELATION OF KW, KVA, AND KVAR

Figure 7-1 shows the triangular relationship that exists between kW, kVA, and kvar. It was pointed out that the vertical line, kvar, was the item we were trying to cancel out with the aid of capacitors. Figure 7-5 illustrates how this component increases with each 10% change of power factor.

Many operators feel that a power factor of 80% on their system is not a serious condition. Even at this so-called "tolerant" condition, the kvar is quite sizable in proportion to the working kilowatts and results in a 25% increase in total kVA in the line. At 70% power factor, this value rises to 42%. However, the kvar at this power factor exceeds the kW, and for each 100 kW of load, 102 kvar of capacitors would be required to cancel out the 102 kvar of the lagging component. This graphic analysis demonstrates the application possibilities. Beneath each of the blocks in Fig. 7-5 is shown the apparent power (kVA) resulting from the combination of the two components.

It is also possible to solve these problems by arithmetic. Suppose that BC of Fig. 7-1 is 3 kvar and that AB is 4 kW. What is AC, the apparent power? In a right triangle, the square of the hypotenuse is equal to the sum of the squares of the other two sides. In this problem, we have

$$(AC)^2 = (3)^2 + (4)^2 = 9 + 16 = 25$$

$$AC = 5 \text{ kVA}$$

From this, we see that the power factor is $\frac{4}{5}$, or .80 or 80%.

The power-factor angle θ, which is measured in degrees, must not be confused with the power factor, which is simply a ratio of kW/kVA

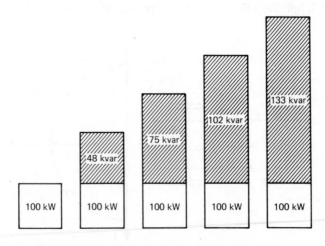

FIGURE 7-5 Graphic relation of kW, kVA, and kvar.

and is expressed as a decimal or in percent. For any given angle, there is but one power factor. For example, when θ is 45°, the power factor is always .707 or 70.7%.

SINE-WAVE ANALYSIS

Sine-wave analysis is another method of determining phase relationship (Fig. 7-6). The horizontal line ABCDE is the baseline, and, from left to right, it represents the passage of time. If A is the present moment, then B, C, D, and E are, progressively, instants of time yet to come. In a 60-Hz system, E will occur $\frac{1}{60}$ sec later than A.

Let us consider curve V, the voltage curve. At A, the voltage is zero; at B, it reaches its maximum positive value and then begins to fall off until at C it is again zero. At D, it has reached its greatest negative value and then begins to rise until at E it is again zero. A fraction of a second after E, it starts through the same cycle of events again so that between A and E the voltage wave has gone through one cycle. Curves I_L and I_C, merely showing the values of current at any given instant of time, are derived in the same manner.

In a 60-Hz system, the time interval between A and E is $\frac{1}{60}$ sec or .0167 sec. In this interval of time, the curves have gone through one cycle, which also has been determined to be 360°. Thus, rather than use cumbersome fractional parts of a second, it is much simpler to express the passage of time by the use of degrees. How much easier to say that C is 180° later than A than to figure the time interval as .0083 sec.

Now consider I_R, the working current. Both I_R and V are at zero, at maximum positive and at maximum negative peaks simultaneously; hence, these two curves are *in phase.*

Why do we mark I_L (inductive or lagging current) on the particular curve selected? At point B, the curve I_L is going through zero and rising, but I_L is doing this 90° later than V; therefore, I_L is lagging V by 90°.

The same reasoning will show that I_C (capacitive current) is leading V by 90°.

The sine-wave analysis is another visual means of determining why capacitive current tends to cancel out inductive current. Anyone who has dabbled in algebra knows that +2 and –2 add up to zero. In Fig. 7-6, the values of I_L and I_C are drawn to the same maximum limits.

Thus, assume that these values were each 10 A at C. However, I_L is +10 and I_C is –10; therefore, their sum is zero. These two currents are equal only at unity (100%) power factor and leave the feeder only the true power represented by the product of voltage V and working current I_R.

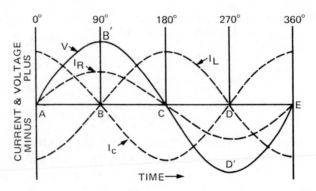

FIGURE 7-6 Sine-wave analysis.

POWER-FACTOR CORRECTION

In actual practice, full correction to establish the unity power factor is rarely, if ever, recommended. If a system had a constant 24-hr load at a given factor, such correction could be readily approached. Unfortunately, such is not the case, and we are faced with peaks and valleys in the load curve.

If we canceled out, by the addition of capacitors, the inductive (lagging) kvars at peak conditions, our capacitors would continuously pump into the system their full value of leading kilovars. Thus, during early morning hours when inductive kvars are much below peak conditions, a surplus of capacitive kvars would be supplied, and a leading power factor would result. Local conditions may justify such overcorrection, but, in general, overcorrection is not recommended. Figure 7-7 will show how far we can go with *fixed* capacitors.

A recording kilovarmeter can readily give us the curve shown, or else it can be calculated if we know the kW curve and the power factor throughout the day. From 2 to 6 a.m., refrigerators, transformer excitation, night factory loads, and the like result in relatively low readings. When the community comes to life in the morning, televisions, radios, appliances, and factory loads build up a high inductive kvar peak.

The area under the shaded section, which is limited by the lowest 24-hr kvar reading, represents the maximum degree of correction to which fixed capacitors are generally applied. It becomes apparent that if we went beyond this point, the utility would be faced with leading power-factor conditions during light-load periods. The shaded area falls into the zone which can be handled effectively with *switched* capacitors.

SWITCHED CAPACITORS

There are several ways in which switching of capacitors can be accomplished. A large factory would arrange by manual or automatic opera-

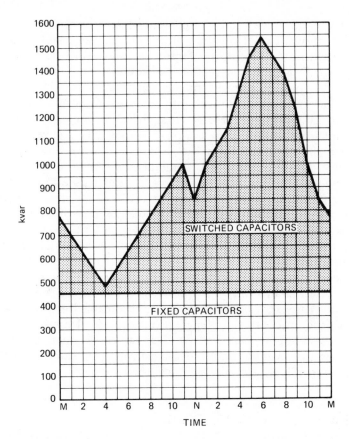

FIGURE 7-7 Graph depicting relation of kvar to time with fixed capacitors.

tion to switch in a bank of capacitors at the start of the working day and disconnect them when the plant shuts down. The energizing of a circuit breaker control coil can be effected with var, current, voltage, temperature, and time controls.

MAKING A SURVEY

The preparation for a survey of system conditions is not as complicated as might first be assumed. Even a very complicated system resolves itself down to a combination of individual feeder studies. Thus, the size of the system does not materially complicate the problem except to increase the amount of manual labor in arriving at the final results.

Assume that Fig. 7-7 represents the kilovar curve for given municipality or for a specific feeder on a large system. Visual inspection of the curve shows that, with the lowest kilovar value being approximately

450 kvar at 4 a.m., we could permanently install on the lines a 450-kvar bank and know that at no time during the day or night would we be operating with a leading power factor. An additional 300-kvar bank or two 300-kvar banks could readily be installed as switched capacitors to correct the power factor in the shaded area.

The larger properties invariably have recording instruments which reveal this kvar demand directly from charts. In the case of the municipal or smaller operator who is not so elaborately equipped, the same results can be obtained from the data available on daily load sheets. Knowledge of the hourly kW demand and the hourly power-factor reading will permit the engineer to readily determine the kvar value at those corresponding times by utilizing available charts. If power-factor readings are not available, kVA can readily be determined from either the armature or the line amps and the system voltage. Once kVA and kW are known, the same chart permits ready determination of power-factor and kvar demand.

Once the total kvar of capacitors required for the system is determined, there remains only the question of proper location. If a severe voltage drop is experienced on the line, the capacitors will serve their best purpose out on the distribution system. However, if voltage problems are not serious on the distribution system and the prime purpose of the installation is to relieve the generators, then the units could very readily be installed at the generating plant.

JUSTIFICATION FOR CAPACITORS

The application of capacitors to electrical distribution systems has been justified by the overall economy provided. Loads are supplied at reduced cost. The original loads on the first distribution systems were predominantly lighting so the power factor was high. Over the years, the character of loads has changed. Today, loads are much larger and consist of many motor-operated devices that impose greater kilovar demands upon electrical systems. Because of the kilovar demand, system power factors have been lower.

The result may be threefold:

1. Substation and transformer equipment may be taxed to full thermal capacity or overburdened.

2. High kilovar demands may, in many cases, cause excessive voltage drops.

3. A low power factor may cause an unnecessary increase in system losses.

Capacitors can alleviate these conditions by reducing the kilovar demand from the point of demand all the way back to the generators. Depending on the uncorrected power factor of the system, the installation of capacitors can increase generator and substation capability for additional load at least 30% and can increase individual circuit capability, from the standpoint of voltage regulation, 30 to 100%.

8

Transformer Construction, Types, and Characteristics

The essential parts of a transformer are shown in Fig. 8-1 and consist of a laminated iron core upon which are wound two separate insulated coils—the primary and the secondary. In most cases, the primary coil is connected to the supply or main side of the line where the alternating current sets up an alternating magnetic flux. This action not only sets up a countervoltage equal and opposite in the primary coil but also sets up a voltage in the secondary coil. The ratio of the voltage in the secondary coil as compared to that in the primary coil depends on the amount of magnetic flux, the frequency of the alternating current, and mainly the number of turns in the coils.

The only current that flows in the primary coil or windings is the magnetizing current necessary to set up the flux in the closed magnetic circuit and is usually a very small percentage of the full-load primary current of the transformer.

TRANSFORMER CONSTRUCTION

To familiarize the reader with the construction of a typical transformer, we shall use a dry-type transformer as manufactured by Hevi-Duty—a unit of General Signal—as an example.

All Hevi-Duty low-voltage, general-purpose transformers rated 15 kVA and larger utilize corrugated aluminum strip coil spacer *ducts* and aluminum strip conductor. This method of coil construction per-

FIGURE 8-1 Three-phase transformer with housing removed to reveal interior parts. Both the primary and secondary connections are made on the insulators located on top of the transformer. Although the division between the primary and secondary coils is not clearly shown, each of the three coils positioned on the bottom of the transformer is divided into two separate parts. (Courtesy R.E. Uptegraff Mfg. Co.)

mits the flow of a large volume of air, and the insulation cannot sag or collapse to reduce or restrict the flow of cooling air.

These large-volume ducts—plus the high heat transfer capability of aluminum strip windings and the heat sink characteristics of the spacers—reduce the transformer's hot-spot temperature rise by an average of 50%. This adds up to cooler operation under all load conditions and a minimum life expectancy of 20 to 25 years.

CORE CONSTRUCTION Many grades of electrical steel are available for use in transformer cores. They fall into two broad classifications: grain-oriented and non-grain-oriented types. Grain-oriented core steel usually exhibits superior magnetic properties over the other type. After being slit to width, it is annealed, edge-coated, and core-plated to assure flatness, lack of burrs, and proper insulation finish. These factors provide for square, tight corners when the laminations are assembled, thereby minimizing core losses and providing quiet operations.

Core laminations are stacked on a special steel frame specifically adjusted for the transformer being assembled. The laminations are carefully aligned to assure proper size, fit, and minimum core losses. The core is then firmly clamped with steel angles, which hold the entire core and coil assembly in place. The clamping angles cover in excess of 90% of the top and bottom of the core to compress all lamination joints and help assure quiet operation.

The core and coil assembly is then preheated in a baking oven to drive off moisture. After drying for a preset time, it is impregnated in varnish and baked until the varnish sets. The varnish gives the core and coil assembly additional mechanical strength and seals it against moisture.

INSULATION SYSTEMS During recent years, the terminology used by electrical equipment manufacturers regarding insulation systems has undergone a major change. Letter designations, such as classes A, B, F, and H are obsolete. Insulation systems are now classified numerically, by temperature rating. What used to be classes A, B, F, and H are now classes 105, 150, 185, and 220, respectively. However, the preceding designations pertain only to the rating of the insulation system. The transformer's rating has also been changed—from classes A, B, F, and H to 55°C rise, 80°C rise, 115°C rise, and 150°C rise. What previously was a class H transformer is now a 150°C rise transformer utilizing a class 220 insulation system. Figure 8-2 shows the old and new designations.

Insulation class	Ambient temperature	Average conductor temperature rise	Hot-spot temperature gradient	Total permissible ultimate temperature
A	40C	55C	10C	105C
B	40C	80C	30C	150C
F	40C	115C	30C	185C
H	40C	150C	30C	220C

FIGURE 8-2 Classes previously identified as A, B, F, and H are now classes 105, 150, 185, and 220, respectively. (Courtesy Acme Corp.)

A well-designed transformer, operating within the temperature rating of its insulation system, will have a life expectancy of 20 to 25 years. The design life of transformers having different insulation systems is the same, and lower-temperature systems will have the same life as higher-temperature systems. The class of insulation used in a particular transformer is a design consideration, and such factors as voltage regulation and material cost and availability are factors that the designer must consider.

TESTING All transformers receive complete electrical testing in accordance with applicable NEMA and ANSI standards. In addition to standard electrical tests, most reputable manufacturers provide for continual monitoring of all major manufacturing processes. This quality control ensures that all transformers measure up to the quality and reliability that users would expect.

TRANSFORMER RATING The rating selected for a transformer installation should handle the normal full load of the system which it feeds, allowing for demand factors. Most transformers will handle as much as 20% of the rated load over long periods of time, but this practice will shorten the life of the transformer as well as increase transformer losses.

If future growth is anticipated, the transformer should be sized for at least its connected load. However, if little future growth is anticipated on the transformer system, then 60% of its connected load is a good rule-of-thumb figure to use in selecting the transformer rating; this, of course, is provided that some diversity can be accounted for.

CHARACTERISTICS OF TRANSFORMERS

In a well-designed transformer, there is very little magnetic leakage. The effect of the leakage is to cause a decrease of secondary voltage when the transformer is loaded. When a current flows through the secondary in phase with the secondary voltage, a corresponding current flows through the primary in addition to the magnetizing current previously mentioned. The magnetizing effects of the two currents are equal and opposite.

In a perfect transformer—one having no eddy-current losses, no resistance in its windings, and no magnetic leakage—the magnetizing effects of the primary load current and the secondary current neutralize each other, leaving only the constant primary magnetizing current effective in setting up the constant flux. If supplied with a constant primary pressure, such a transformer would maintain constant secondary pressure at all loads. Obviously, the perfect transformer has yet to be built; the closest is a transformer with very small eddy-current loss where the drop in pressure in the secondary windings is not more than 1 to 3% depending on the size of the transformer.

INSULATING TRANSFORMERS

Insulating transformers are those with no internal electrical connection between the primary and the secondary windings. In all but a few designs, it is customary to first wind the low-voltage coil next to the core and, once completed, finish the unit by winding the high-voltage coil over it. This construction places the conductors energized at the high voltage a greater physical distance from the magnetic core, which is normally grounded. The core is electrically interconnected with core clamps, steel structure, and enclosing case, all of which are connected with a ground lead to the plant or system ground.

In the usual high-voltage transformer design, the lower voltage coil is wound next to the core. This in turn is covered with mica sheet layer insulation over which the high-voltage coil is wound. Additional sheet mica insulation is applied around each coil with a final wrap of glass tape for extra electrical and mechanical strength. This is commonly known as *barrel*-type construction with a 220°C insulation system.

Cooling ducts are strategically placed within each winding to carry away the internally generated heat. The smooth exterior coil surfaces in a vertical plane minimize the accumulation of dirt.

TAPS

If an electric utility could always guarantee to deliver exactly the rated primary voltage at every transformer location, taps would be unnecessary. However, it is not possible to achieve this, and in recognition of this fact, the public service commissions of the various states allow reasonable variations above or below a nominal value.

Generally speaking, if a load is very close to a substation or generating plant, the voltage will consistently be above normal. Near the end of the line the voltage may be below normal. The primary taps are used to match these voltages.

In large transformers, it would be very inconvenient to move the thick, well-insulated primary leads to different tap positions when changes in source-voltage levels make this desirable. Therefore, taps are used, such as shown in the wiring diagram in Fig. 8-3. In this transformer, the permanent high-voltage leads would be connected to H1 and H2 and the secondary leads, in their normal fashion, to X1, X2, X3, and X4. Note, however, the tap arrangements available at 2, 3, 4, 5, 6, and 7. Until a pair of these taps is interconnected with a jumper, the

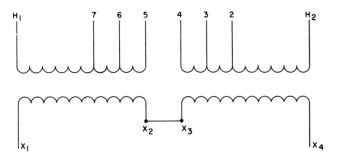

FIGURE 8-3 Diagram of a typical transformer showing tap connections. (Courtesy Sorgel Co.)

primary circuit is not completed. If this were the typical 7200-V primary, for example, the transformer would have a normal 1620 turns. Assume 810 of these turns are between H1 and 6 and another 810 between 3 and H2. Then if we connect 6 and 3 together with a flexible jumper on which lugs have already been installed, the primary circuit is completed, and we have a normal ratio transformer that could deliver 120/240 V from the secondary.

Between tap 6 and either 5 or 7, we have the familiar 40 turns. Similarly, between 3 and either 2 or 4, we also have 40 turns. From what you have learned previously, changing the jumper from 3 to 6 to 3 to 7 removes 40 turns from the left half of the primary. The same condition would apply on the right half of the winding if the jumper were between 6 and 2. Either connection would boost secondary voltage by $2\frac{1}{2}\%$. Had we connected 2 and 7, 80 turns would have been omitted and a 5% boost resulted. Placing the jumper between 6 and 4 or 3 and 5 would reduce output voltage by 5%.

9

Transformer Connections— Polarity

The polarity of a transformer is determined by the lead markings. See Fig. 9-1. When X_1 is located diagonally with respect to H_1, the polarity is, by definition, additive. When H_1 and X_1 are adjacent, the polarity is said to be subtractive.

One of the most interesting transformer connections is the Scott scheme which was originated by C.F. Scott to take care of the necessity of changing two-phase power generated at Niagara Falls to three-phase power for transmission to Buffalo, N.Y. This goes back to 1896 when Mr. Scott was Chief Electrician of the Westinghouse Electric Corp.

Since that early day there have been other connections, as will be shown, to take care of this transformation. Since the best of these schemes requires three transformers as compared to two in the Scott system, none has proved of great practical importance.

It is well to remember in making emergency connections, especially for phase transformation, that a comparatively small *off ratio* will cause a large circulating current. This circulating current will depend not only on the off ratio but also on the impedance. This condition may become more dangerous than the operator realizes, especially when the transformers involved are normally carrying full load.

TRANSFORMER POLARITY

Polarity of a transformer is an indication of direction of flow of current from a terminal at any one instant. The idea is quite similar to the polarity marking on a battery.

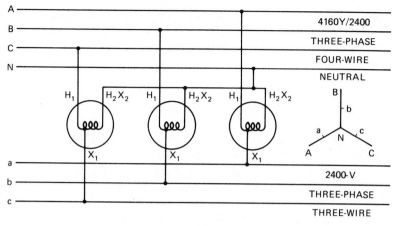

(a)

SINGLE-PHASE

(b)

FIGURE 9-1 (a) Wiring diagram showing lead markings which determine the polarity of a transformer. (b) Single-phase, 120 V transformer connection.

As you face the high-voltage side of a transformer, the high-voltage terminal on your right is always marked H_1 (H-1) and the other high-voltage terminal is marked H_2 (H-2). This is an established standard.

In making transformer connections—particularly bank connections—the polarity of individual transformers must be checked. In making such connections it is necessary to remember that all H_1 terminals are of the same polarity and all X_1 terminals are of the same polarity. So if you were to connect two single-phase transformers in parallel, you should connect the two H_1 terminals together, then the two H_2 terminals together, the two X_1 terminals together, and the two X_2 transformer terminals together. By following this procedure, you can satisfactorily parallel transformers regardless of whether they are both of the same polarity or whether one is additive and one is subtractive polarity.

The ANSI standards for transformers specify additive polarity as standard on all single-phase units in sizes 200 kVA and smaller having high-voltage windings 8600 V and below. All other units are subtractive polarity.

TRANSFORMER CONNECTIONS, SINGLE PHASE

FOR LIGHTS AND 120-V POWER   In the early days of residential electrical service, it was quite common to supply only a 30-A, 120-V single-phase service, as lighting was the primary load with the occasional use of small 120-V appliances such as a radio, toaster, iron, and the like. For this type of service, a transformer was placed between the high-voltage line and the line containing low voltage, as shown in Fig. 9-1. The 120/240-V low-voltage windings are connected in parallel, giving 120 V on a two-wire system.

FOR 240-V POWER ONLY   The connection, shown in Fig. 9-2, is similar to the one shown previously except that the windings are connected in series to produce 240 V on a two-wire system.

FOR LIGHT AND POWER   The connection shown in Fig. 9-3 is the most common single-phase distribution system in use today. It is known as the three-wire, 240/120-V single-phase system and is used where 120 and 240 V are used simultaneously.

TWO-PHASE SYSTEM   Two single-phase circuits operating 90° out of phase with each other are called a two-phase system. Two different transformer connections are normally used. The one shown in Fig. 9-4 is treated as two separate circuits, while the connection in Fig. 9-5 has a common wire on the secondary side, resulting in some saving

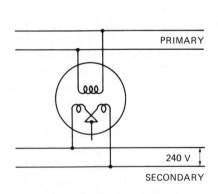

FIGURE 9-2 Transformer connection giving 240 V on a two-wire system.

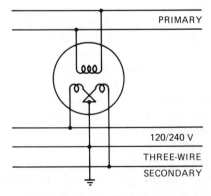

FIGURE 9-3 The most common transformer connection for light and power.

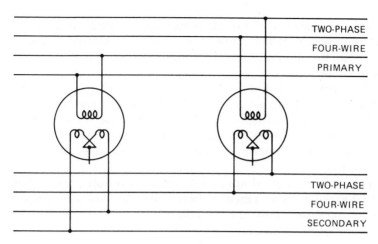

FIGURE 9-4 Two-phase, four-wire secondary system.

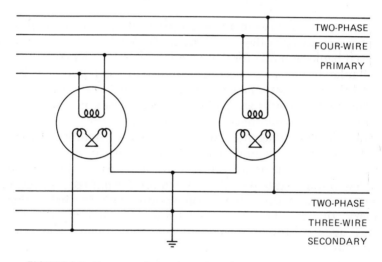

FIGURE 9-5 Common wire used on a two-phase system to save copper.

in copper. In the latter connection, the common wire must carry 1.41 × current in other secondary wires.

THREE-PHASE CONNECTIONS

Any combination of additive and subtractive units can be connected in three-phase banks so long as the correct polarity relationship of terminals is observed. Whether a transformer is additive or subtractive does not alter the designation of the terminals; thus, the correct polarity will be assured if connections are made as indicated in the diagrams. The terminal designations, if not marked, can be obtained from the transformer nameplate, which shows the schematic internal-connection diagrams with the actual physical relationship between the high- and the low-voltage terminals.

If subtractive-polarity transformers are used in three-phase banks, secondary connections are simplified from those shown for the additive-polarity units. The additive-polarity connections, for standard angular displacement, are somewhat complicated, particularly in cases with a delta-connected secondary, by the crossed secondary interconnections between units.

For this reason, simplified bank connections, which give non-standard angular displacement between primary and secondary systems, are sometimes used with additive-polarity units.

STANDARD ANGULAR DISPLACEMENT

Standard angular displacement or vector relationships between the primary and secondary voltage systems are 0° for delta-delta- or wye-wye-connected banks and 30° for delta-wye or wye-delta banks.

Angular displacement becomes important when two or more three-phase banks are interconnected into the same secondary system or when three-phase banks are parallel. In such cases, it is necessary that all the three-phase banks have the same displacement.

The following diagrams cover three-phase circuits using standard connections—where all units have additive polarity and give standard angular displacement or a vector relation between the primary and secondary voltage systems. Also shown are simplified connections for the more common three-phase connections with the delta-connected secondary, that is, where all units have additive polarity but give non-standard angular displacement between the primary and secondary voltage systems.

DELTA-DELTA FOR POWER   The transformer connection shown in Fig. 9-6 has been one of the most commonly used connections of the past. The ungrounded primary system will continue to supply power even though one of the lines is grounded due to a fault. If one of the transformers in the bank should fail, secondary power can be supplied from the two remaining units by changing to the open delta connection. Thus, this type of connection is ideal from the standpoint of service continuity.

The connection in Fig. 9-7 is similar to the one in Fig. 9-6 except that the former gives a 180° angular displacement, which is nonstandard. Otherwise the information given for the one in Fig. 9-6 is applicable.

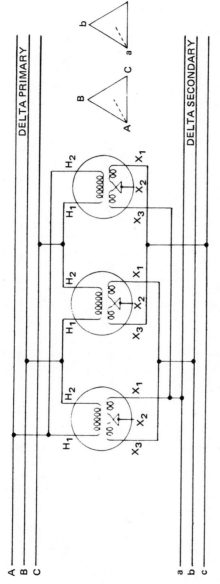

FIGURE 9-6   Delta-delta connection for power.

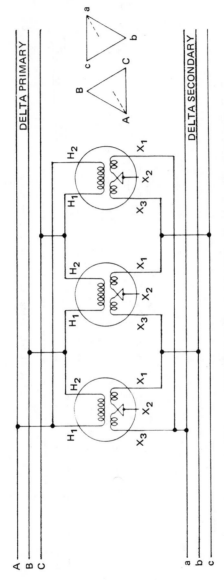

FIGURE 9-7 Simplified connection for the power connection in Fig. 9-6.

126

DELTA-DELTA FOR LIGHT AND POWER When light and power are to be supplied from the same bank of transformers, the midtap of the secondary of one of the transformers is grounded and connected to the fourth wire of the three-phase secondary system, as shown in Fig. 9-8. The lighting load is then divided between the two hot wires of this same transformer, the grounded wire being common to both branches.

A simplified connection for light and power is shown in Fig. 9-9.

OPEN DELTA FOR POWER The circuit connection shown in Fig. 9-10 can be used in an emergency in case one of the transformers in a delta-delta bank fails. This type of bank is also used to supply power to a three-phase load which is temporarily light but which is expected to grow. When the load increases to a point where the two transformers in the bank are overloaded, an increase in capacity of 1.732 times can be obtained by adding another unit of the same size and using the delta-delta connection.

The capacity of an open delta bank is only 57.7% of a delta-delta bank of the same sized units.

A similar connection to the one shown in Fig. 9-10 is shown in Fig. 9-11, but the one in Fig. 9-11 gives a 180° angular displacement, which is nonstandard.

OPEN DELTA FOR LIGHT AND POWER When the secondary circuits are to supply both light and power, the open delta bank takes the form shown in Fig. 9-12. In addition to the applications listed for the open delta bank for power, this type of bank is used where there is a large single-phase load and only a small three-phase load. In this case, the two transformers would be of different kVA sizes, the one across which the lighting load is connected being the larger. This is also the connection that should be used when protected transformers are employed in a three-phase bank supplying both light and power.

A simplified connection of an open delta for light and power is shown in Fig. 9-13.

Y-DELTA FOR POWER The present tendency in utilities is to replace the 2400 delta system with the 2400/4160Y-V three-phase four-wire system. This change in effect raises the distribution voltage from 2400 to 4160 V without any major changes in connected equipment. The same transformers that were previously connected between lines on the 2400-V delta system are now connected between lines and neutral on the new 2400/4160Y-V system. Three-phase banks that had previously been connected delta-delta are now connected Y-delta, as shown in Fig. 9-14. A simplified connection of this system is shown in Fig. 9-15.

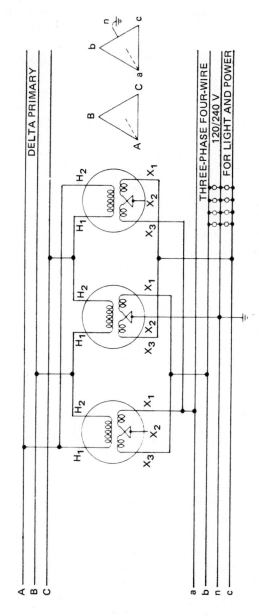

FIGURE 9-8 Delta-delta connection for light and power.

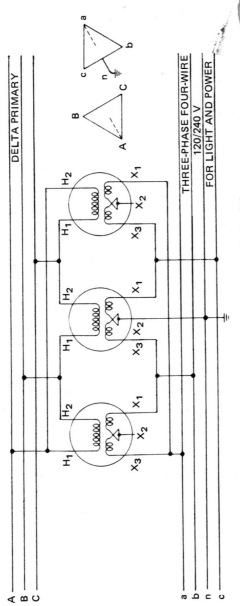

FIGURE 9-9 Simplified connection for Fig. 9-8.

129

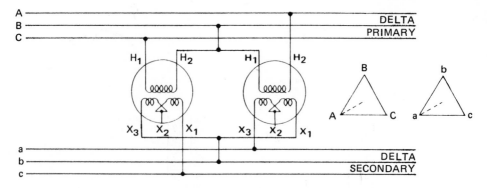

FIGURE 9-10 Open delta connection for power.

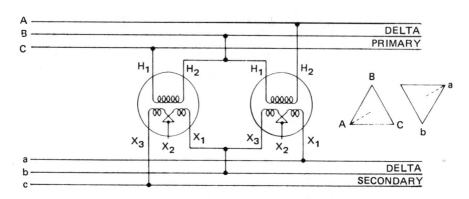

FIGURE 9-11 Simplified connection for power connection shown in Fig. 9-10.

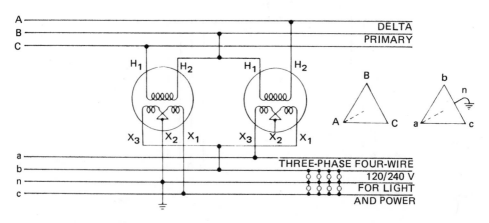

FIGURE 9-12 Open delta connection for light and power.

130

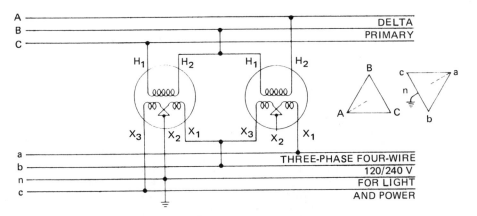

FIGURE 9-13 Simplified connection for Fig. 9-12.

Y-DELTA FOR LIGHT AND POWER When service for both light and power is needed, the Y-delta bank takes the form shown in fig. 9-16. Note that the secondary connections are the same as for the delta-delta bank. A simplified connection is shown in Fig. 9-17.

Y-DELTA WITH ONE UNIT MISSING If one unit of a Y-delta bank goes bad, service can be maintained by means of the connection shown in Fig. 9-18. In the regular Y-delta bank with three units, the neutral of the primaries of the transformers is not ordinarily tied in with the neutral of the primary system. In fact, this bank can be used even when the primary neutral is not available. In the bank with two units, however, it is necessary to connect to the neutral, as shown in the diagram. The main disadvantage of this connection is the fact that full-load current flows in the neutral even though the three-phase load may be balanced. In addition to maintaining service in an emergency, this type of bank is satisfactory where the main part of the load is lighting and the three-phase load is small. This connection is for additive-polarity units and gives standard angular displacement.

DELTA-Y FOR LIGHT AND POWER In all the connections mentioned previously in which both power and lighting are served on the secondary, the grounded secondary wire is not the neutral of the three-phase system but rather the midpoint of one leg of the delta. Furthermore, all the lighting load is put on one phase. The primary currents in any one bank, therefore, are unbalanced.

To obtain balanced currents at the substation or generator, the transformers supplying the lighting loads are connected to different primary phases in the different three-phase banks. This means that different legs of the delta are grounded in the various three-phase banks. This type of secondary connection can, therefore, not be used when the

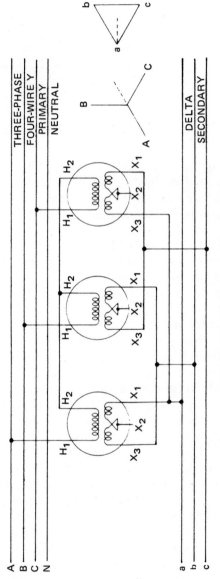

FIGURE 9-14 Y-delta connection for power.

132

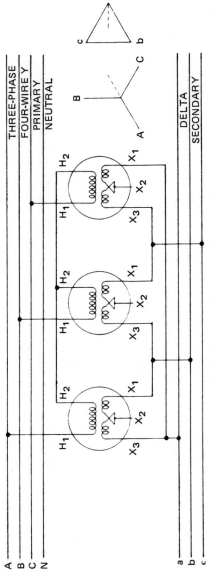

FIGURE 9-15 Simplified connection for Fig. 9-14.

133

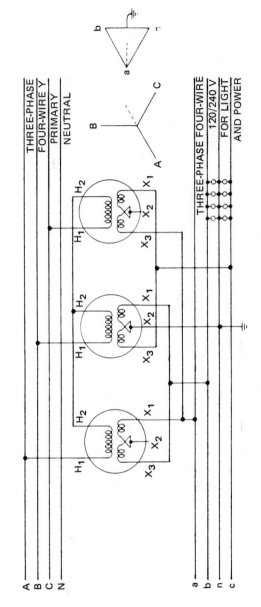

FIGURE 9-16 Y-delta connection for light and power.

134

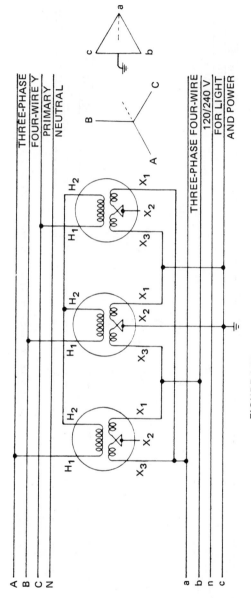

FIGURE 9-17 Simplified connection for Fig. 9-16.

135

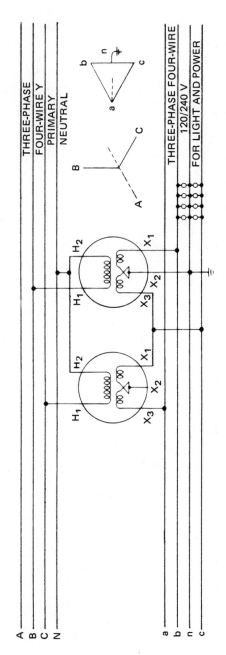

FIGURE 9-18   Y-delta connection with one unit missing.

136

secondaries of different banks are to be paralleled or banked. In regions of high load density, superior operating characteristics are obtained when secondaries are banked. The delta-Y connection is most commonly used when this is done. See Fig. 9-19. Note that standard transformers with 120/240-V secondary windings are used with the secondary of each unit connected for 120 V. In this case, the neutral of the secondary three-phase system is grounded. The single-phase loads are connected between different phase wires and neutral, while the three-phase power loads are connected to the three-phase wires. Thus, 120 V is supplied so that the secondaries of different banks can be tied together. This connection is for additive-polarity units and gives standard angular displacement.

Y-Y FOR LIGHT AND POWER The primaries of the transformers can also be connected Y, as shown in Fig. 9-20. When the primary system is 2400/4160Y V, 2400-V transformers would be used in place of the 4160-V transformers that would be required for the delta-Y connection. A saving in transformer cost results. It is necessary that the primary neutral be available when this connection is used, and the neutrals of the primary system and of the bank are tied together as shown. If the three-phase load is unbalanced, part of the load current flows in the primary neutral. For these reasons, it is very necessary that the neutrals be tied together as shown. If this tie were omitted, the line to neutral voltages on the secondary would be very unstable. That is, if the load on one phase were heavier than on the other two, the voltage on this phase would drop excessively and the voltage on the other two phases would rise. Also, large third-harmonic voltages would appear between lines and neutral, both in the transformers and in the secondary system in addition to the 60-Hz component of voltage. This means that for a given value of rms voltage, the peak voltate would be much higher than for a pure 60-Hz voltage. This overstresses the insulation both in the transformers and in all apparatus connected to the secondaries.

Y-Y AUTOTRANSFORMERS FOR SUPPLYING POWER In the changeover from the 2400-V delta to the 2400/4160Y primary system mentioned previously, there may be some customers on a utility line with 2400-V motors which cannot be connected directly to the new primary system. The cheapest way of adapting these motors to the new system is by means of a bank of autotransformers connected as shown in Fig. 9-21. In this application, the size and cost of autotransformers are considerably less than for two winding transformers delivering the same amount of power, due to the fact that the ratio of transformation is small.

SCOTT CONNECTION—THREE PHASE TO TWO PHASE Many old distribution systems which originally were two phase are being

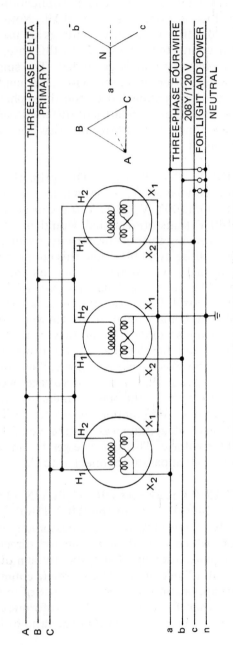

FIGURE 9-19 Delta-Y connection for light and power.

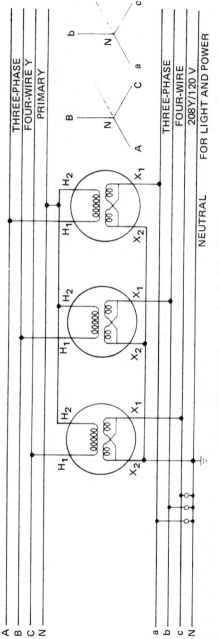

FIGURE 9-20 Y-Y connection for light and power.

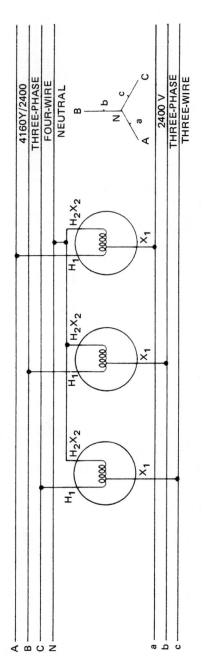

FIGURE 9-21 Y-Y autotransformer connection for supplying power to 2400-Volt motors from a 2400/4160Y-V three-phase four-wire system.

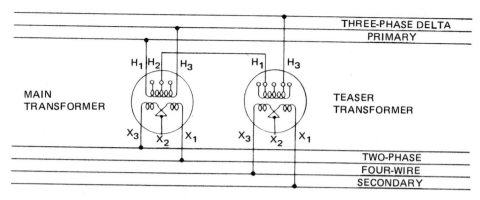

FIGURE 9-22 Scott connection—three phase to two phase.

changed over to three phase. Utilities making this changeover are faced with the problem of supplying power to customers having two-phase motors. The Scott bank is most generally used in this connection. The transformers are special in that they have taps at 50 and 86.6% in the primary winding. The currents in the primary windings of these transformers are $15\frac{1}{2}\%$ greater than when transforming single phase at the same kVA and voltage. This means that a transformer with Scott taps built on the same core as a regular single-phase transformer will have only 86.6% of the kVA capacity of the single-phase transformer. See Fig. 9-22.

SCOTT CONNECTION—TWO PHASE TO THREE PHASE In the case of a factory being built in a district that is at present served by a two-phase system, three-phase rather than two-phase motors would probably be installed in the factory, since they are more standard and since there is a good chance of the two-phase system being changed over to three phase at some future date. Here the utility company is faced with the problem of supplying three-phase power from a two-phase system. The Scott connection with three phase on the secondary performs this function. In this case, the currents in the secondary windings are $15\frac{1}{2}\%$ greater than in transformers delivering the same kVA single phase at the same voltage. This again causes the kVA output of transformers with Scott taps in the secondary to be 86.6% of regular transformers of the same physical size. This connection is for additive-polarity units and gives standard angular displacement. See Fig. 9-23.

SCOTT CONNECTION—THREE PHASE TO THREE PHASE As two-phase systems are gradually changed over to three phase, it is not necessary to replace the old two-phase banks but just to reconnect them. The teaser transformer will operate with 86.6% voltage on both the primary and secondary, so it should have the same ratio as the main transformer. See Fig. 9-24.

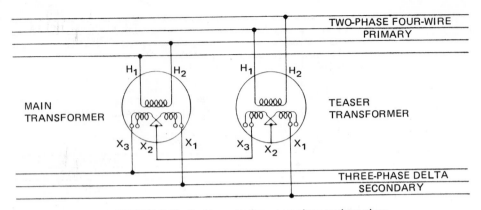

FIGURE 9-23   Scott connection – two phase to three phase.

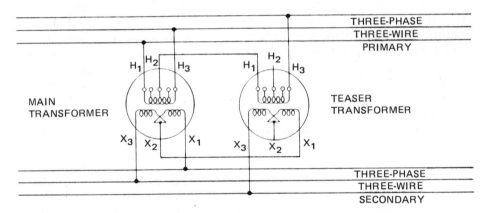

FIGURE 9-24   Scott connection – three phase to three phase.

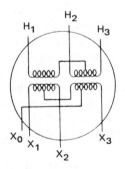

FIGURE 9-25   Modern version of the Scott T connection.

The modern version of this Scott connection is in a duplex transformer, called a T connection, which can simulate the various wye or delta connections except that no neutral will be made available on the primary side. These T-connected units will have a 30° angular displacement and may have a neutral on the secondary side. See Fig. 9-25.

10

Parallel Operation
of Transformers

Transformers will operate satisfactorily in parallel (Fig. 10-1) on a single-phase, three-wire system if the terminals with the same relative polarity are connected together. This is really not a very economical operation because the individual cost and losses of the smaller transformers are greater than one larger unit giving the same output. In other words, paralleling of smaller transformers is usually done only in an emergency. In large transformers, however, it is often practical to operate units in parallel as a regular practice.

In connecting large transformers in parallel, especially when one of the windings is for a comparatively low voltage, the resistance of the joints and interconnecting leads must not vary materially for the different transformers, or it will cause an unequal division of load.

Two three-phase transformers may also be connected in parallel provided they have the same winding arrangement, are connected with the same polarity, and have the same phase rotation. If two transformers—or two banks of transformers—have the same voltage ratings, the same turn ratios, the same impedances, and the same ratios of reactance to resistance, they will divide the load current in proportion to their kVA ratings, with no phase difference between the currents in the two transformers. However, if any of the preceding conditions are not met, then it is possible for the load current to divide between the two transformers in proportion to their kVA ratings. There may also be a phase difference between currents in the two transformers or banks of transformers.

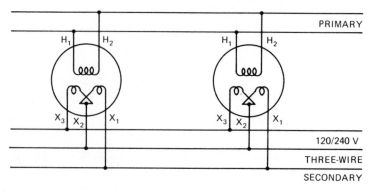

FIGURE 10-1 Two single-phase transformers can be connected in parallel if the terminals with the same relative polarity are connected together.

Some three-phase transformers cannot be operated properly in parallel. For example, a transformer having its coils connected in delta on both high-tension and low-tension sides cannot be made to parallel with one connected either in delta on the high-tension and in Y on the low-tension or in Y on the high-tension and in delta on the low-tension side. However, a transformer connected in delta on the high-tension side and in Y on the low-tension side can be made to parallel with transformers having their coils joined in accordance with certain schemes, connected in star of Y on the high-tension side and in delta on the low-tension side.

To determine whether or not three-phase transformers will operate in parallel, connect them as shown in Fig. 10-2, leaving two leads on one of the transformers unjoined. Test with a voltmeter across the unjoined leads. If there is no voltage between the points shown in the

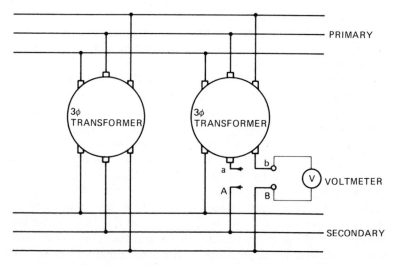

FIGURE 10-2 Testing three-phase transformers for parallel operation.

drawing, the polarities of the two transformers are the same, and the connections may then be made and put into service.

If a reading indicates a voltage between the points indicated in the drawing (either one of the two or both), the polarity of the two transformers are different. Should this occur, disconnect transformer lead A successively to mains 1, 2, and 3 as shown in Fig. 10-2 and at each connection test with the voltmeter between b and B and the legs of the main to which lead A is connected. If with any trial connection the voltmeter readings between b and B and either of the two legs is found to be zero, the transformer will operate with leads b and B connected to those two legs. If no system of connections can be discovered that will satisfy this condition, the transformer will not operate in parallel without changes in its internal connections, and there is a possibility that it will not operate in parallel at all.

In parallel operation, the primaries of the two or more transformers involved are connected together, and the secondaries are also connected together. With the primaries so connected, the voltages in both primaries and secondaries will be in certain directions. It is necessary that the secondaries be so connected that the voltage from one secondary line to the other will be in the same direction through both transformers. Proper connections to obtain this condition for single-phase transformers of various polarities are shown in Fig. 10-3. In Fig.

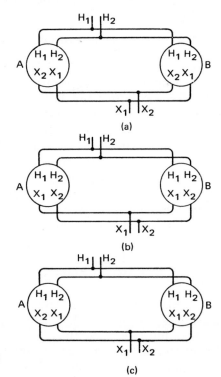

FIGURE 10-3 Transformers connected in parallel.

10-3(a), both transformers A and B have additive polarity; in Fig. 10-3(b), both transformers have subtractive polarity; in Fig.10-3(c), transformer A has additive polarity and B has subtractive polarity.

Transformers, even when properly connected, will not operate satisfactorily in parallel unless their transformation ratios are very close to being equal and their impedance voltage drops are also approximately equal. A difference in transformation ratios will cause a circulating current to flow, even at no load, in each winding of both transformers. In a loaded parallel bank of two transformers of equal capacities, for example, if there is a difference in the transformation ratios, the load circuit will be superimposed on the circulating current. The result in such a case is that in one transformer the total circulating current will be added to the load current, whereas in the other transformer the actual current will be the difference between the load current and the circulating current. This may lead to unsatisfactory operation. Therefore, the transformation ratios of transformers for parallel operation must be definitely known.

When two transformers are connected in parallel, the circulating current caused by the difference in the ratios of the two is equal to the difference in open-circuit voltage divided by the sum of the transformer impedances, because the current is circulated through the windings of both transformers due to this voltage difference. To illustrate, let I represent the amount of circulating current—in percent of full-load current—and the equation will be

$$I = \frac{\text{percent voltage difference} \times 100}{\text{sum of percent impedances}}$$

Let's assume an open-circuit voltage difference of 3% between two transformers connected in parallel. If each transformer has an impedance of 5%, the circulating current, in percent of full-load current, is $I = (3 \times 100)/(5 + 5) = 30\%$. A current equal to 30% full-load current therefore circulates in both the high-voltage and low-voltage windings. This current adds to the load current in the transformer having the higher induced voltage and subtracts from the load current of the other transformer. Therefore, one transformer will be overloaded, while the other may or may not be—depending on the phase-angle difference between the circulating current and the load current.

IMPEDANCE IN PARALLEL-OPERATED TRANSFORMERS

Impedance plays an important role in the successful operation of transformers connected in parallel. The impedance of the two or more transformers must be such that the voltage drop from no load to full load is the same in all transformer units in both magnitude and phase. In most applications, you will find that the total resistance drop is relatively small when compared with the reactance drop and that the total percent impedance drop can be taken as approximately equal to the percent reactance drop. If the percent impedances of the given transformers at full load are the same, they will of course divide the load equally.

The following equation may be used to obtain the division of loads between two transformer banks operating in parallel on single-phase systems. In this equation, it can be assumed that the ratio of resistance to reactance is the same in all units since the error introduced by differences in this ratio is usually so small as to be negligible:

$$\text{power} = \frac{(kVA - 1)/(Z - 1)}{[(kVA - 1)/(Z - 1)] + [(kVA - 2)/(Z - 2)]} \times \text{total kVA load}$$

where

$$kVA - 1 = kVA \text{ rating of transformer 1}$$

$$kVA - 2 = kVA \text{ rating of transformer 2}$$

$$Z - 1 = \text{percent impedance of transformer 1}$$

$$Z - 2 = \text{percent impedance of transformer 2}$$

The preceding equation may also be applied to more than two transformers operated in parallel by adding, to the denominator of the fraction, the kVA of each additional transformer divided by its percent impedance.

PARALLEL OPERATION OF THREE-PHASE TRANSFORMERS

Three-phase transformers, or banks of single-phase transformers, may be connected in parallel provided each of the three primary leads in one three-phase transformer is connected in parallel with a corresponding primary lead of the other transformer. The secondaries are then connected in the same way. The corresponding leads are the leads which have the same potential at all times and the same polarity. Furthermore, the transformers must have the same voltage ratio and the same impedance voltage drop.

When three-phase transformer banks operate in parallel and the three units in each bank are similar, the division of the load can be determined by the same method previously described for single-phase transformers connected in parallel on a single-phase system.

In addition to the requirements of polarity, ratio, and impedance, paralleling of three-phase transformers also requires that the angular displacement between the voltages in the windings be taken into consideration when they are connected together.

Phasor diagrams of three-phase transformers that are to be paralleled greatly simplify matters. With these, all that is required is to compare the two diagrams to make sure they consist of phasors that can be made to coincide; then connect together terminals corresponding to coinciding voltage phasors. If the diagram phasors can be made to coincide, leads that are connected together will have the same potential at all times. This is one of the fundamental requirements for paralleling.

PHASOR DIAGRAMS

When three single-phase transformers are used as a three-phase bank, the direction of the voltage in each of the six phase windings may be represented by a voltage phasor. A voltage phasor diagram of the six voltages involved provides a convenient way to study the relative direction and amounts of the primary and the secondary voltages.

The same is true for one three-phase transformer, which also has six voltages to be considered, because its phase windings on the high-voltage side and the low-voltage side are connected together in the same way as the phase windings of three single-phase transformers.

To show how a phasor diagram is drawn, consider the three-phase transformer shown in Fig. 10-4. Here we have a Y-delta-connected three-phase transformer with three legs and each carrying a high- and low-voltage winding. The high-voltage windings are connected in Y (with a common neutral point at N) with the leads to the high-voltage terminals designated H_1, H_2, and H_3. The three low-voltage windings are connected in delta. The junction points of the three windings serve as low-voltage terminals X_1, X_2, and X_3.

The voltages in the low-voltage windings are assumed to be equal to each other in amount but are displaced from each other 120°.

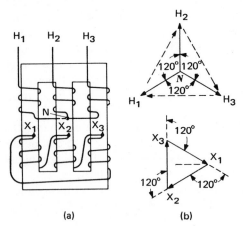

(a)   (b)   FIGURE 10-4 Windings and phasor diagrams of three-phase transformer.

When drawing the phasor diagram for the low-voltage windings, phasor X_1X_2 is drawn first in any selected direction and to any convenient scale. The arrowhead indicates the instantaneous direction of the alternating voltage in winding X_1X_2, and the length of the phasor represents the amount of voltage in the winding. The broken lines that extend past the arrowheads represent reference lines for phase angles. Since winding X_2X_3 is physically connected to the end of X_2 in the X_1X_2 winding, phasor X_2X_3 will be started at point X_2, 120° out of phase with phasor X_1X_2 in the clockwise direction. The length of phasor X_2X_3 is equal to the length of phasor X_1X_2. Winding X_3X_1 is drawn in a similar manner.

The high-voltage phasor comes next. Since the low-voltage winding X_1X_2 is wound on the same leg as the high-voltage winding NH_1, the voltages in these two windings are in phase and are represented by parallel voltage phasors. Therefore, the phasor NH_1 is drawn from a selected point N parallel to X_1X_2. Note that all three high-voltage windings are physically connected to a common point N. Therefore, the phasors representing the voltages NH_1, NH_2, and NH_3 will all start at the common point N in the phasor diagram. Again, the high-voltage phasors are all of the same length but displaced from each other by 120° as shown.

The high-voltage phasors have the same length as the low-voltage phasors. Each phase of the high voltage is, however, proportional to the low voltage in the same phase according to the turns ratio, making the voltage in each phase of the high-voltage winding higher than those of the low-voltage windings. Since the high-voltage windings are connected in Y, each line voltage is 1.732 times higher than a phase voltage in any one winding. It can be geometrically proved, for example, that $H_1H_2 \times NH_1 = 1.732 \times NH_2$ and so forth for each line voltage.

In comparing three-phase transformers for possible operation in parallel, draw the voltage phasor diagram for transformer bank A and mark the terminals as shown in Fig. 10-5. Next draw the phasor diagram for bank B on drafting or other transparent paper; draw a heavy

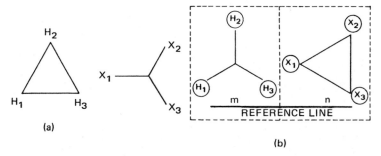

(a)

(b)

FIGURE 10-5 Phasor diagrams for three-phase transformer banks in parallel.

reference line m-n as shown in the drawing in Fig. 10-5. Cut the transparent diagram into two parts as indicated by the dashed lines in the drawing. The two parts of the reference line are now marked m and n. Place diagram m on the high-voltage diagram of (a) so that the terminals which are desired to be connected together coincide. Place diagram n on the low-voltage diagram of (a) so that the heavy reference line of n is parallel to the heavy reference line of m. If, under these conditions, the terminals of n can be made to coincide with the low-voltage terminals of (a), the terminals which coincide can be connected together for parallel operation. If the low-voltage terminals cannot be made to coincide, parallel operation is not possible with the assumed high-voltage connection.

11

Connections
and Applications
of Autotransformers

The basic difference of an autotransformer and a double-wound transformer is that an autotransformer is a transformer whose primary and secondary circuits have part of a winding in common and therefore the two circuits are not isolated from each other. The application of an autotransformer is a good choice for some users where a 480Y/277- or 208Y/120-V, three-phase, four-wire distribution system is utilized. The main advantages are as follows:

1. Lower purchase price
2. Lower operating cost due to lower losses
3. Smaller size; easier to install
4. Better voltage regulation
5. Lower sound levels

An autotransformer, however, cannot be used on a 480- or 240-V, three-phase, three-wire delta system. A grounded neutral phase conductor must be available in accordance with NE Code Article 210-9, which is as follows:

210-9. Circuits Derived from Autotransformers. Branch circuits shall not be supplied by autotransformers. Exception No. 1: Where the system supplied has a grounded conductor that is electrically connected to a grounded conductor of the system supplying the autotransformer.

The NE Code, in general, requires that separately derived alternating current systems be grounded. The secondary of a two-winding, insulated transformer is a separately derived system. Therefore, it must be grounded in accordance with NE Code Article 250-26 as follows:

250-26. Grounding Separately Derived Alternating/Current Systems. A separately derived ac system that is required to be grounded by Section 250-5 shall be grounded as specified in (a) through (d) below.

(a) Bonding Jumper. A bonding jumper, sized in accordance with Section 250-79(c) for the derived phase conductors, shall be used to connect the equipment grounding conductors of the derived system to the grounded conductor. Except as permitted by Exception No. 4 of Section 250-23(a), this connection shall be made at any point on the separately derived system from the source to the first system disconnecting means or overcurrent device; or it shall be made at the source of a separately derived system which has no disconnecting means or overcurrent devices.

Exception: The size of the bonding jumper for a system that supplies a Class 1 remote control or signaling circuit, and is derived from a transformer rated not more than 1000 volt-amperes, shall not be smaller than the derived phase conductors and shall not be smaller than No. 14 copper or No 12 aluminum wire.

(b) Grounding Electrode Conductor. A grounding electrode conductor, sized in accordance with Section 250-94 for the derived phase conductors, shall be used to connect the grounded conductor of the derived system to the grounding electrode as specified in (c) below. Except as permitted by Exception No. 4 of Section 250-23(a), this connection shall be made at any point on the separately derived system from the source to the first system disconnecting means or overcurrent device; or it shall be made at the source of a separately derived system which has no disconnecting means or overcurrent devices.

Exception: A grounding electrode conductor shall not be required for a system that supplies a Class 1 remote control or signaling circuit, and is derived from a transformer rated not more than 1000 volt-amperes, provided the system grounded conductor is connected to the transformer frame or enclosure by a jumper sized in accordance with the Exception for (a), above, and the transformer frame or enclosure is grounded by one of the means specified in Section 250-57.

(c) Grounding Electrode. The grounding electrode shall be as near as practicable to and preferably in the same area as the grounding conductor connection to the system. The grounding electrode shall be: (1) the nearest available effectively grounded structural metal member of the structure; or (2) the nearest available effectively grounded metal water pipe; or (3) other electrodes as specified in Sections 250-81 and 250-83 where electrodes specified by (1) or (2) above are not available.

(d) Grounding Methods. In all other respects, grounding methods shall comply with requirements prescribed in other parts of this Code.

A typical drawing for the above NE Code requirement is shown in Fig. 11-1. In the case of an autotransformer, the grounded conductor of

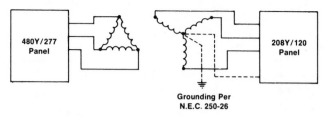

FIGURE 11-1   Grounding connection per NE Code 250-26.

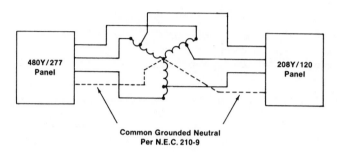

FIGURE 11-2   Autotransformers are grounded as shown here. This complies with NE Code 210-9.

the supply is brought into the transformer to the common H0-X0 terminal, and the ground is established to satisfy the NE Code. See Fig. 11-2.

PRACTICAL APPLICATION: AUTOTRANSFORMER FOR 208Y/120-V SERVICE

Many buildings are presently served with 480Y/277 V for their main distribution system. However, these same structures also require 208Y/ 120 V for convenience outlets, computer installations, and the like. The conventional way to provide this service is to use small step-down transformers. Figure 11-3 shows a typical application of this system. Only the secondary winding of the large power transformer is shown. Between the power transformer and the two sets of windings to the right is a representation of the plant bus system to which numerous installations such as an insulating-type transformer could be connected. The delta primary served at 480 V is transformed to a wye-connected secondary from which 120 V is obtained for convenience outlets is connected to X1, X2, or X3, to X0. The same results can be achieved, at a somewhat lower cost, by utilizing an autotransformer.

In Fig. 11-4, it can be seen that the only change from Fig. 11-3 is in the right-hand position of the diagram. Note that instead of a delta-wye insulating transformer an autotransformer is used.

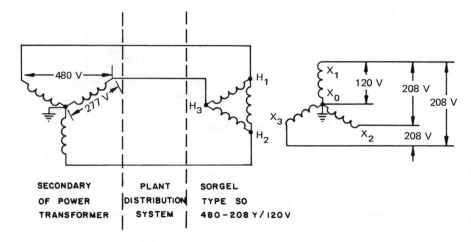

FIGURE 11-3 Typical application using small step-down transformers.

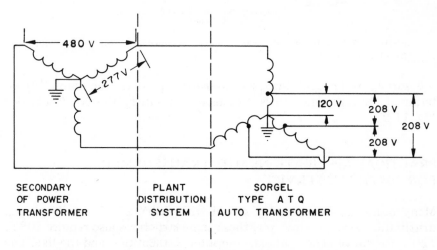

FIGURE 11-4 Same application as in Fig. 11-3 except an autotransformer is used in place of the delta-wye insulating transformer.

In Fig. 11-5(a), a conventional 50 kVA insulating transformer designed with a 277-V primary and a 120-V secondary is shown. If 50,000 V-A (50 kVA) is divided by 277 V, the primary current calculates to be 180 A. Assume the load to be 50 kVA. On the secondary side, the current would be $50,000/120 = 416.0$ A if transformer losses are ignored.

With the primary voltage from X to Z equal to 277 V, it should be obvious that at some point Y, a voltmeter could read 120 V from Y to Z. This would leave 157 V from X to Y. Between Y_1 and Z_1, we also know the design potential is 120 V, so there should be no harm in connecting Y to Y_1 and Z to Z_1. If this were done, it would produce the

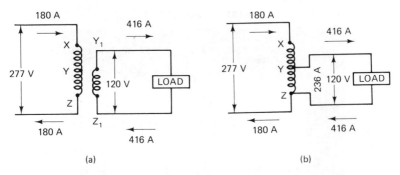

FIGURE 11-5 (a) A conventional 50 kVA insulating transformer designed with a 277-V primary and a 120-V secondary. (b) Diagram explaining the operation of an autotransformer.

circuit shown in Fig. 11-5(b). This is exactly what is done in an auto-transformer.

Referring again to Fig. 11-5(b), because the load is still 50,000 V-A, the secondary current will remain 416 A. Similarly, the primary must supply 50 kVA; hence the primary lines continue to feed in 180 A. Note, however, that a change has taken place in the winding between Y and Z. It can be seen that 416 A flows in the secondary wire to the left toward Z but only 180 A flows away from Z toward the 277-V source. This is the difference of 236 A. Since it is impossible to store amperes anywhere, all current arriving at a point such as Z must leave that point. The only place it can flow is upward from Z to Y. With 236 A coming up from Z and 180 A going down from X, the sum of these at Y must move to the right as 416 A of secondary current, balancing out completely.

In Fig. 11-5(a), the secondary winding has to be rugged enough to carry 416 A, but in Fig. 11-5(b), that winding could be made of smaller wire because the current is now only 236 A. This reduces the cost of copper, the size of coil, and the weight. A further saving in weight appears in the smaller core that is required to serve this new electrical configuration. This also reduces exciting current.

A performance data comparison between Sorgel Watchdog[R] auto-transformers and standard two-winding insulated transformers is shown in Fig. 11-6. In using autotransformers, however, be aware of the following reminders:

1. *Isolation systems for computers or other sensitive loads.* Since the source and the load share a common winding, there is no isolation of the load from the source. Where this is a requirement, it is recommended that shielded isolating transformers be used.

2. *Short-circuit currents.* From the table in Fig. 11-6, it can be seen that the impedances of the autotransformers are quite a bit lower than comparable two-winding transformers. Although this provides better voltage regulation, it must be considered when making short-circuit studies for proper protective devices.

AVERAGE INDUSTRY DATA

KVA	Autotransformer 480Y/277-208Y/120 3 Phase, 60 Hertz 150°C Rise	Insulating Transformer 480 Delta—208Y/120 3 Phase, 60 Hertz 150°C Rise
30	Core— .5% Load— 1.2% Total— 1.7% Full Load Efficiency—98.3%	Core— .7% Load— 4.6% Total— 5.3% Full Load Efficiency—94.7%
45	Core— .5% Load— 2.3% Total— 2.8% Full Load Efficiency—97.2%	Core— .6% Load— 4.4% Total— 5.0% Full Load Efficiency—95.0%
75	Core— .4% Load— 2.4% Total— 2.8% Full Load Efficiency—97.2%	Core— .6% Load— 3.3% Total— 3.9% Full Load Efficiency—96.1%
112	Core— .4% Load— 1.7% Total— 2.1% Full Load Efficiency—97.9%	Core— .4% Load— 3.3% Total— 3.7% Full Load Efficiency—96.3%
150	Core— .3% Load— 2.0% Total— 2.3% Full Load Efficiency—97.7%	Core— .4% Load— 3.0% Total— 3.4% Full Load Efficiency—96.6%
225	Core— .3% Load— 1.7% Total— 2.0% Full Load Efficiency—98.0%	Core— .4% Load— 3.0% Total— 3.4% Full Load Efficiency—96.6%
300	Core— .2% Load— 1.3% Total— 1.5% Full Load Efficiency—98.5%	Core— .4% Load— 1.8% Total— 2.2% Full Load Efficiency—97.8%

All losses based on operating temperature of 150°C rise plus 20°C ambient = 170°C reference temperature in accordance with NEMA, ANSI and IEEE standards.

IMPEDANCE AT 170°C

KVA	Autotransformers	Insulating Transformers
30	2.1%	6.4%
45	3.3%	6.6%
75	3.7%	5.7%
112	2.4%	6.1%
150	3.5%	5.5%
225	2.6%	6.6%
300	3.5%	3.6%

SOUND LEVELS

KVA	Design Levels		NEMA Standard
	Auto	Insulating	
30	44 DB	43 DB	45 DB
45	44 DB	44 DB	45 DB
75	44 DB	47 DB	50 DB
112	44 DB	49 DB	50 DB
150	50 DB	50 DB	50 DB
225	50 DB	51 DB	55 DB
300	52 DB	54 DB	55 DB

FIGURE 11-6 Performance data comparison between autotransformers and two-winding insulated transformers.

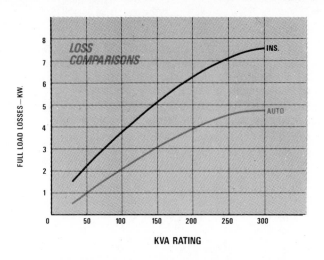

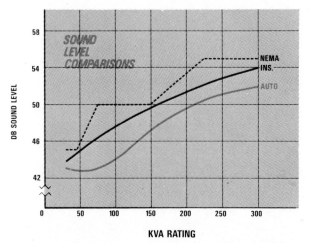

TWO WINDING INSULATING TRANSFORMER ───────
AUTOTRANSFORMER
NEMA STD - - - - - - - - -

FIGURE 11-6 (cont.)

3. Again, remember to use an autotransformer only on three-phase, four-wire systems, where the grounded neutral conductor is present. The transformer is completely interchangeable to either step-down from a 480Y/277-V system to 208Y/120 V or to step-up from 208Y/120 V to 480Y/277 V.

4. *Third harmonics.* These are always present in wye-wye connections but are kept at a minimum by using a three-legged core construction. For general use, this presents no problems.

12

Control Transformers

The term *control transformers* could be used to describe a large variety of transformers designed for many different purposes. However, for our use, control transformer will be defined as a device used to reduce supply voltages to 120 V or lower for the operation of electromagnetic devices such as contactors, solenoids, relays, and the like.

Industrial control transformers are especially designed to accommodate the momentary current inrush caused when electromagnetic components are energized—without sacrificing secondary voltage stability beyond practical limits.

Most control transformers are dry-type step-down units with the secondary control circuit isolated from the primary line circuit to assure maximum safety. These transformers and other components are usually mounted within an enclosed control box or control panel which has push-button stations independently grounded as recommended by the NE Code and other safety codes.

Other types of control transformers—sometimes referred to as *control and signal* transformers—normally do not have the required industrial control transformer regulation characteristics. Rather, they are constant-potential, self-air-cooled transformers used for the purpose of supplying the proper voltage (usually reduced) for control circuits of electrically operated switches or other equipment and, of course, for signal circuits. Some are of the open type with no protective casing over the windings, while others are enclosed with a metal casing over the winding.

Bell transformers, as the name implies, are small-capacity constant-potential transformers used for operating doorbells, buzzers, door openers, and annunciators. They are self-cooled units usually enclosed in metal cases. They are manufactured for use on primary voltages of 120 or 240 V (sometimes 277 V) with a single secondary voltage of about 10 V or with multiple taps giving secondary voltages of 6, 12, 18, and 24 V.

Instrument transformers are also a form of control transformer that are employed for the purpose of stepping down the voltage of a circuit for the operation of instruments and gauges. Instrument transformers are usually of two types: potential and current.

POTENTIAL TRANSFORMER In general, a potential transformer supplies low voltage to an instrument which is connected to its secondary. The voltage is proportional to the primary voltage, but it is small enough to be safe for the test instrument. The secondary of a potential transformer may be designed for several different voltages, but most are designed for 120 V.

As may well be imagined, these transformers are used with devices requiring voltage for operation, such as voltmeters, frequency meters, power-factor meters, watt-hour meters, and the like. In the case of a multimeter, one transformer may be used for any number of instruments at the same time, provided the total current usage does not exceed the rating of the potential transformer.

The potential transformer is primarily a distribution transformer especially designed for good voltage regulation, so that the secondary voltage under all conditions will be as nearly as possible a definite percentage of the primary voltage.

CURRENT TRANSFORMER An instrument transformer is so called because it is normally used to supply current or voltage of a smaller value than the line current or voltage to an electrical instrument.

A current transformer supplies current to an instrument connected to its secondary, the current being proportional to the primary current but small enough to be safe for the instrument. The secondary of a current transformer is usually designed for a rated current of 5 A.

A current transformer operates in the same way as any other transformer; that is, the same relations exist between the primary and the secondary current and voltages. A current transformer is connected in series with the power lines to which it is applied, so that line current flows in its primary winding. The secondary of the current transformer is connected to current devices such as ammeters, wattmeters, watt-hour meters, power-factor meters, some forms of relays, and trip coils of some types of circuit breakers.

When no instruments or other devices are connected to the secondary of the current transformer, a short-circuit device or connection is placed across the secondary. In other words, the secondary circuit of a current transformer should never be opened while the primary is carrying current. Before disconnecting an instrument, the secondary of the current transformer must be short-circuited. If the secondary circuit is opened while the primary winding is carrying current, there will be no secondary ampere turns to balance the primary ampere turns, so the total primary current becomes exciting current and magnetizes the core to a high flux density, which produces a high voltage across both primary and secondary windings. Since, to secure accuracy, current transformers are designed with normal exciting currents of only a small percentage of full-load current, the voltage produced with the secondary open-circuited is high enough to endanger the life of anyone coming in contact with the meters or leads. Also, the high secondary voltage may overstress the secondary insulation and cause a breakdown. Operation with the secondary open-circuited may also cause the transformer core to become permanently magnetized. If this should occur, the core may be demagnetized by passing about 50% excess current through the primary, with the secondary connected to an adjustable high resistance that is gradually reduced to zero.

SELECTING CONTROL TRANSFORMERS

Even though control and signal transformers are relatively small in size and power, the loads must still be calculated and completely analyzed before the proper transformer selection can be made. This analysis involves every electrically energized component in the control circuit.

All electromagnetic control devices have two current requirements: the first to energize the coil and the second to maintain the contact for a definite period of time. The initial energizing of the coil—which takes 5 to 20 sec—requires many times more current than normal. This is referred to as volt-ampere inrush, which is immediately followed by the amount of current required to hold the contact in the circuit and is referred to as sealed volt-amperes.

To select an appropriate control transformer for a given application, first determine the voltage and frequency of the supply circuit. Then determine the total inrush VA of the control circuit. In doing so, do not neglect the current requirements of indicating lights and timing devices that do not have an inrush VA but are energized at the same time as the other components in the circuit. Their total VA should be added to the total inrush VA.

Refer to the regulation data chart in Fig. 12-1. If the supply

REGULATION DATA CHART

Continuous Nominal VA	Inrush VA at 20% PF		
(NAMEPLATE RATING)	95% Sec. Voltage	90% Sec. Voltage	85% Sec. Voltage
50	200	240	280
75	350	470	580
100	400	575	770
150	800	950	1250
250	1500	2200	2750
300	2000	2800	3900
350	3200	3700	4900
500	4200	5800	8000
750	8000	11000	15000
1000	13000	18000	23000
1500	15000	24000	31000
2000	20000	32000	41000
3000	39000	60000	77000
5000	75000	120000	150000

FIGURE 12-1   Regulation data chart.

circuit voltage (as determined previously) is reasonably stable and fluctuates no more than ±5%, refer to the 90% secondary voltage column. However, should it fluctuate as much as 10%, refer to the 95% secondary voltage column. Whichever column applies, go down it until you arrive at the inrush VA closest to—but not less than—the inrush VA of the control circuit in question.

Continue by reading to the far left side of the chart and you will have selected the continuous nominal VA rating of the transformer needed. The secondary voltage that will be delivered under inrush conditions will be either 85, 90, or 95% of the rated secondary voltage, depending on the column selected from the regulation data chart in Fig. 12-1. Also refer to the table in Fig. 12-2 for typical magnetic motor starter and contactor data.

Refer to the specification tables found in manufacturers' catalogs to select a transformer according to the required continuous nominal VA and primary and secondary voltages. A few typical ones are shown in Figs. 12-3 through 12-7. A table of recommended ratings for slow blow fuses is shown in Fig. 12-8.

TYPICAL MAGNETIC MOTOR STARTER & CONTACTOR DATA
60 Hz, 120 Volt, 3-Pole

CONTACTOR MANUFACTURER	N.E.M.A. SIZE							
	00	0	1	2	3	4	5	
500 SERIES		192	192	240	660	1225		VA Inrush
ALLEN BRADLEY		29	29	29	45	69		VA Sealed
700 SERIES	53	110	175	240	580	1000	1945	VA Inrush
	15	20	22	31	43	65	98	VA Sealed
ASEA			75	230	400	900	1600	VA Inrush
			10	30	45	55	100	VA Sealed
CROUSE-HINDS (ARROW HART DIV.)	175	175	175	175	680	680	5500	VA Inrush
	21	21	21	21	45	45	240	VA Sealed
CUTLER-HAMMER (CITATION LINE)	87	104	104	394	1034	1034	1034	VA Inrush
	15	20	20	51	100	100	100	VA Sealed
FURNAS	218	218	218	218	440	957	1518	VA Inrush
	25	25	25	25	45	75	116	VA Sealed
GENERAL ELECTRIC	144	144	144	528	1152	1248	3600	VA Inrush
	24	24	24	60	83	86	276	VA Sealed
GTE-SYLVANIA	75	120	200	550	1140	1380	2700	VA Inrush
	20	30	35	80	120	145	370	VA Sealed
ITE GOULD		198	198	360	790	1400	900	VA Inrush
		24	24	41	57	70	10	VA Sealed
SQUARE D	118	245	245	311	700	1185	2970	VA Inrush
	11	27	27	37	46	85	212	VA Sealed
WESTINGHOUSE	160	160	160	160	625	625		VA Inrush
	25	25	25	25	50	50		VA Sealed

FIGURE 12-2   Typical magnetic motor starter and contactor data.

U.L. RECOGNIZED COMPONENT   CSA LISTED

PRIMARY VOLTS 240/480/600, 230/460/575, 220/440/550			SECONDARY VOLTS 120/100, 115/95, 110/90				50/60 Hz		
CATALOG NO.	VA RATING	OUTPUT AMPS	A	B	C	D	E	F	WT. IN LBS.
TA-1-81000	50	.435	4 1/4	3 1/16	2 11/16	2 1/2	2 1/2	3/16 x 3/8	3 3/4
TA-1-81009	75	.652	4 7/16	3 1/16	2 11/16	2 1/2	2 11/16	3/16 x 3/8	4 1/4
TA-1-81001	100	.870	4 7/8	3 1/16	2 11/16	2 1/2	3 1/8	3/16 x 3/8	4 1/2
TA-1-81002	150	1.305	4 11/16	3 7/8	3 5/16	3 1/8	3 1/16	7/32 x 1/2	6 3/4
TA-1-81003	250	2.175	6 1/4	4 9/16	3 15/16	3 3/4	3 1/2	7/32 x 1/2	11
TA-1-81020	300	2.61	6 13/16	4 15/16	4 1/4	4 1/16	4 1/16	7/32 x 1/2	15
TA-1-81004	350	3.045	6 13/16	4 15/16	4 1/4	4 1/16	4 1/16	7/32 x 1/2	15
TA-1-81005	500	4.35	7 5/16	5 5/16	4 9/16	4 3/8	4 9/16	5/16 x 1/2	20
TA-1-81006	750	6.51	7 5/16	6 13/16	5 13/16	5 3/4	3 13/16	5/16 x 1/2	25
TA-1-81007	1000	8.70	8 1/8	6 13/16	5 13/16	5 3/4	4 5/8	5/16 x 1/2	31
TA-1-81008	1500	13.05	9 7/8	6 13/16	5 13/16	5 3/4	6 3/8	5/16 x 1/2	47
TA-1-53929	2000	17.4	8	7 9/16	7 3/4	6 1/2	6 5/8	13/32 x 13/16	53
TA-1-53930	3000	26.1	8 3/4	10 15/16	9 3/8	7 3/4	5 1/2	7/32 x 13/16	75
TA-1-53931	5000	43.5	10 3/8	11 15/16	10 1/8	7 3/4	6	7/32 x 13/16	112

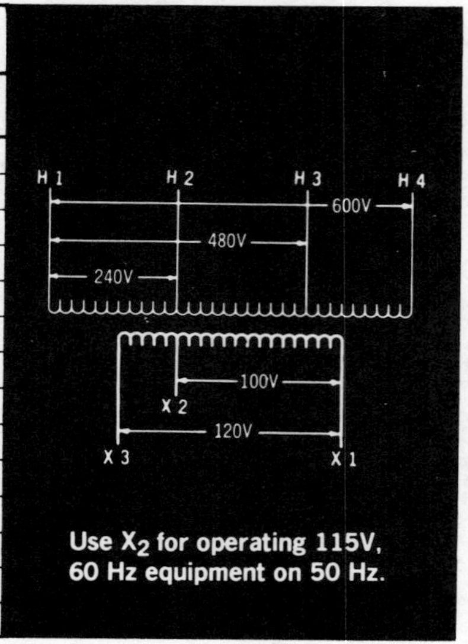

H 1
H 2
H 3
H 4
600V
480V
240V
100V
X 2
120V
X 3
X 1
Use X_2 for operating 115V, 60 Hz equipment on 50 Hz.

FIGURE 12-3 Specifications for control transformers with secondary voltage from 90 to 120 V.

U.L. RECOGNIZED COMPONENT CSA LISTED

PRIMARY VOLTS 208/240/277 380/480 SECONDARY VOLTS 24 50/60 Hz

CATALOG NO.	VA RATING	OUTPUT AMPS	A	B	C	D	E	F	WT. IN LBS.
TA-1-81321	50	2.08	3 13/16	3 13/16	3 5/16	3 1/8	2 3/16	7/32 x 1/2	4
TA-1-81322	75	3.12	3 7/8	3 13/16	3 5/16	3 1/8	2 1/4	7/32 x 1/2	4 3/4
TA-1-81323	100	4.17	4 1/4	3 13/16	3 5/16	3 1/8	2 5/8	7/32 x 1/2	5 1/2
TA-1-81324	150	6.25	5 3/4	4 9/16	3 15/16	3 3/4	3	7/32 x 1/2	7
TA-1-81325	250	10.4	6	4 15/16	4 1/4	4 1/16	3 1/4	7/32 x 1/2	11
TA-1-81326	350	14.6	6 3/4	5 5/16	4 9/16	4 3/8	4	5/16 x 1/2	16 1/2
TA-1-81327	500	20.8	6 15/16	5 5/16	4 9/16	4 3/8	4 3/16	5/16 x 1/2	19
TA-1-81328	750	31.2	7 5/16	6 13/16	5 13/16	5 3/8	3 13/16	5/16 x 1/2	26
TA-1-81329	1000	41.7	8 1/8	6 13/16	5 13/16	5 3/4	4 5/8	5/16 x 1/2	33

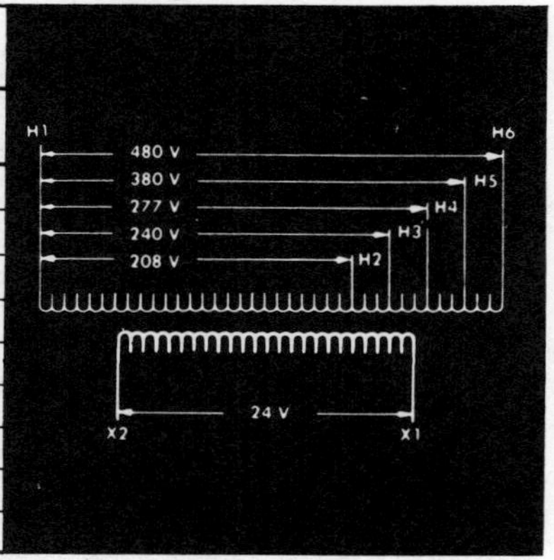

FIGURE 12-4 Specifications for 24-V (secondary) control transformers.

U.L. RECOGNIZED COMPONENT — CSA LISTED

PRIMARY VOLTS 208/277/380				SECONDARY VOLTS 115/95				50/60 Hz	
CATALOG NO.	VA RATING	OUTPUT AMPS	A	B	C	D	E	F	WT. IN LBS.
TA-1-81301	50	.435	$4\frac{3}{16}$	$3\frac{1}{16}$	$2\frac{11}{16}$	$2\frac{1}{2}$	$2\frac{7}{16}$	$\frac{3}{16} \times \frac{3}{8}$	$3\frac{3}{4}$
TA-1-81302	75	.652	$4\frac{1}{2}$	$3\frac{1}{16}$	$2\frac{11}{16}$	$2\frac{1}{2}$	$2\frac{3}{4}$	$\frac{3}{16} \times \frac{3}{8}$	$4\frac{1}{4}$
TA-1-81303	100	.870	$4\frac{1}{16}$	$3\frac{13}{16}$	$3\frac{5}{16}$	$3\frac{1}{8}$	$2\frac{7}{16}$	$\frac{7}{32} \times \frac{1}{2}$	$4\frac{1}{2}$
TA-1-81304	150	1.305	$4\frac{11}{16}$	$3\frac{13}{16}$	$3\frac{5}{16}$	$3\frac{1}{8}$	$3\frac{1}{16}$	$\frac{7}{32} \times \frac{1}{2}$	7
TA-1-81305	250	2.175	$6\frac{3}{4}$	$4\frac{9}{16}$	$3\frac{15}{16}$	$3\frac{3}{4}$	4	$\frac{7}{32} \times \frac{1}{2}$	12
TA-1-81306	350	3.045	$7\frac{1}{2}$	$4\frac{15}{16}$	$4\frac{1}{4}$	$4\frac{1}{16}$	$4\frac{3}{4}$	$\frac{7}{32} \times \frac{1}{2}$	18
TA-1-81307	500	4.35	$7\frac{5}{16}$	$5\frac{5}{16}$	$4\frac{9}{16}$	$4\frac{3}{8}$	$4\frac{9}{16}$	$\frac{5}{16} \times \frac{1}{2}$	19
TA-1-81308	750	6.51	$7\frac{3}{16}$	$6\frac{13}{16}$	$5\frac{13}{16}$	$5\frac{3}{4}$	$3\frac{11}{16}$	$\frac{5}{16} \times \frac{1}{2}$	23
TA-1-81309	1000	8.7	$7\frac{15}{16}$	$6\frac{13}{16}$	$5\frac{13}{16}$	$5\frac{3}{4}$	$4\frac{7}{16}$	$\frac{5}{16} \times \frac{1}{2}$	$33\frac{1}{4}$

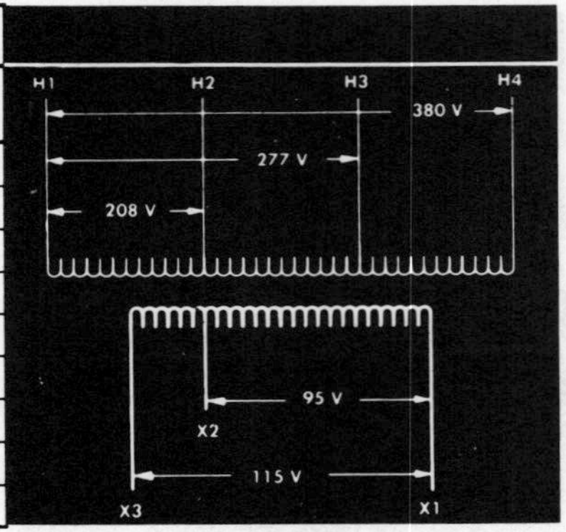

FIGURE 12-5 Specifications for 208/277/380-primary-volt to 115/95-secondary-volt instrument transformers.

U.L. RECOGNIZED COMPONENT CSA LISTED

PRIMARY VOLTS 600/550; 440/380		SECONDARY VOLTS 230/115							50/60 Hz	
CATALOG NO.	VA RATING	OUTPUT AMPS 115V	230V	A	B	C	D	E	F	WT. IN LBS.
TA-1-54535	50	.435	.218	$3\frac{13}{16}$	$3\frac{13}{16}$	$3\frac{5}{16}$	$3\frac{1}{8}$	$2\frac{3}{16}$	$\frac{7}{32}$ x $\frac{1}{2}$	4
TA-1-54536	100	.870	.435	$3\frac{15}{16}$	$3\frac{13}{16}$	$3\frac{5}{16}$	$3\frac{1}{8}$	$2\frac{5}{16}$	$\frac{7}{32}$ x $\frac{1}{2}$	5
TA-1-54537	150	1.305	.653	$4\frac{11}{16}$	$3\frac{13}{16}$	$3\frac{5}{16}$	$3\frac{1}{8}$	$3\frac{1}{16}$	$\frac{7}{32}$ x $\frac{1}{2}$	7
TA-1-54538	250	2.17	1.09	$6\frac{1}{4}$	$4\frac{9}{16}$	$3\frac{15}{16}$	$3\frac{3}{4}$	$3\frac{1}{2}$	$\frac{7}{32}$ x $\frac{1}{2}$	11
TA-1-81197	350	3.045	1.523	$7\frac{1}{8}$	$4\frac{15}{16}$	$4\frac{1}{4}$	$4\frac{1}{16}$	$4\frac{3}{8}$	$\frac{7}{32}$ x $\frac{1}{2}$	13
TA-1-54539	500	4.35	2.17	$8\frac{1}{2}$	$4\frac{15}{16}$	$4\frac{1}{4}$	$4\frac{1}{16}$	$5\frac{3}{4}$	$\frac{7}{32}$ x $\frac{1}{2}$	23
TA-1-81240	750	6.51	3.26	$7\frac{3}{16}$	$6\frac{13}{16}$	$5\frac{13}{16}$	$5\frac{3}{4}$	$3\frac{11}{16}$	$\frac{5}{16}$ x $\frac{1}{2}$	25
TA-1-81241	1000	8.70	4.35	$7\frac{15}{16}$	$6\frac{13}{16}$	$5\frac{13}{16}$	$5\frac{3}{4}$	$4\frac{7}{16}$	$\frac{5}{16}$ x $\frac{1}{2}$	31

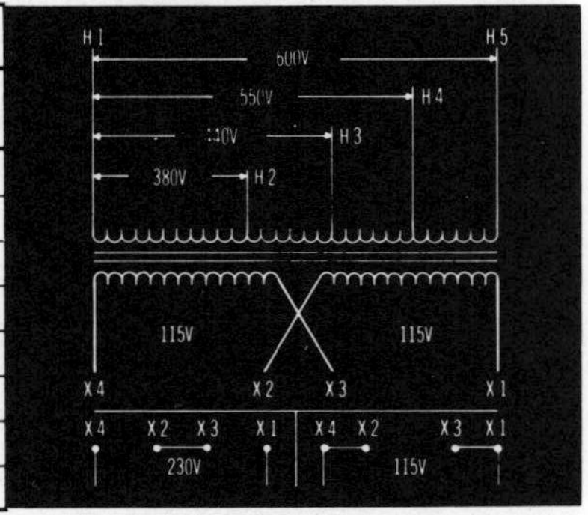

FIGURE 12-6 Specifications for 600/550- and 440/380-primary-volt to 230/115-secondary-volt instrument transformers.

UNIVERSAL VOLTAGE U.L. RECOGNIZED COMPONENT CSA LISTED

PRIMARY VOLTS 208/220/230/240; 380/400/416/440/460/480; 500/550/575/600 — SECONDARY VOLTS 85/91/95/99; 100/110/115/120; 125/130 — 50/60 Hz

CATALOG NO.	VA RATING	OUTPUT AMPS*	A	B	C	D	E	F	WT. IN LBS.
TA-1-32403	50	.435	3 7/8	3 13/16	3 5/16	3 1/8	2 1/4	7/32 x 1/2	4 3/8
TA-1-32404	150	1.305	5 3/4	4 9/16	3 15/16	3 3/4	3	7/32 x 1/2	10
TA-1-32405	250	2.175	6 13/16	4 15/16	4 1/4	4 1/16	4 1/16	7/32 x 1/2	14
TA-1-32669	350	3.045	8 1/4	4 15/16	4 1/4	4 1/16	5 1/2	7/32 x 1/2	16 5/8
TA-1-32406	500	4.35	7 1/8	6 13/16	5 13/16	5 3/4	3 5/8	5/16 x 1/2	21 7/8
TA-1-54523	750	6.51	7 5/8	6 13/16	5 13/16	5 3/4	4 1/8	5/16 x 1/2	29
TA-1-54524	1000	8.70	8 3/8	6 13/16	5 13/16	5 3/4	4 7/8	5/16 x 1/2	32
TA-1-54525	1500	13.05	7 5/8	7 9/16	7 3/4	6 1/2	6 1/4	13/32 x 13/16	51
TA-1-81202	2000	17.4	8	7 9/16	7 3/4	6 1/2	6 5/8	13/32 x 13/16	55
TA-1-81203	3000	26.1	8 7/8	10 15/16	9 3/8	7 3/4	5 1/2	13/32 x 13/16	75
TA-1-81205	5000	43.5	11 3/4	11 15/16	10 1/4	7 3/4	6	13/32 x 13/16	114

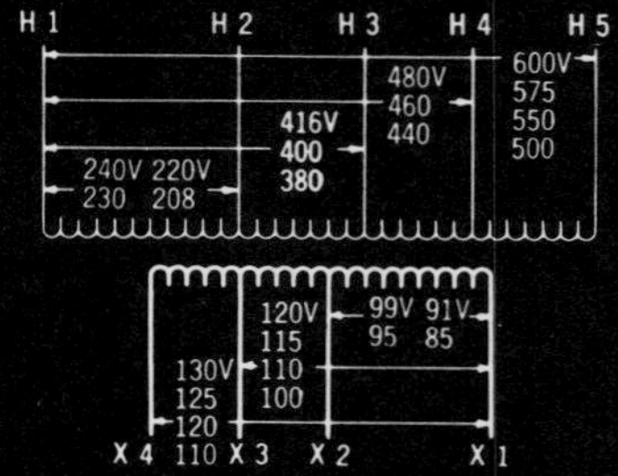

Only X2 or X3 may be used at this Amp rating. Both X2 and X3 may be used at the same time provided total amps do not exceed figure shown.

FIGURE 12-7 Instrument transformer specifications for units capable of a wide voltage range on both primary and secondary sides.

VA	230 VOLT SECONDARY			115 VOLT SECONDARY			24 VOLT SECONDARY		
	FULL LOAD SEC AMPS	FUSE AMPS	FUSE KIT PART NUMBER	FULL LOAD SEC AMPS	FUSE AMPS	FUSE KIT PART NUMBER	FULL LOAD SEC AMPS	FUSE AMPS	FUSE KIT PART NUMBER
50	.217	3/10	PL-112600	.435	1/2	PL-112600	2.08	2-1/2	PL-112600
75	.326	4/10	PL-112600	.652	8/10	PL-112600	3.13	3-2/10	PL-112600
100	.435	1/2	PL-112600	.870	1	PL-112600	4.17	5	PL-112600
150	.652	8/10	PL-112600	1.30	1-1/2	*PL-112600	6.25	7	*PL-112600
250	1.09	1-1/8	PL-112601	2.17	2-1/4	PL-112601	10.42	15	PL-112601
300	1.30	1-4/10	PL-112601	2.61	2-8/10	PL-112601	12.5	15	PL-112601
350	1.52	1-6/10	PL-112601	3.04	3-2/10	PL-112601	14.58	20	PL-112601
500	2.17	2-1/4	PL-112601	4.35	4-1/2	PL-112601	20.83	25	PL-112601
750	3.26	3-1/2	PL-112601	6.52	7	PL-112601			
1000	4.35	4-1/2	PL-112601	8.70	9	PL-112601			
1500	6.52	7	PL-112601	13.40	15	PL-112601			
2000	8.70	9	PL-112601						

FIGURE 12-8   Recommended rating for slow blow fuses.

169

AIR CONDITIONING, REFRIGERATION, AND APPLIANCE TRANSFORMERS

The transformers falling under this category are autotransformers designed to change a wide range of voltages to the standard motor voltages for domestic appliances, air conditioners, and related equipment. It is advantageous to correct any high or low supply voltage conditions to match the voltage requirements of the appliances and equipment in order to assure safe, efficient operation.

Special autotransformers will change or correct off-standard voltage that may be the result of the following:

1. Line supply voltage which does not match the appliance motor nameplate voltage (see the schematic diagram in Fig. 12-9).
2. Low voltage due to inadequate capacity of wiring in the electrical distribution system.
3. Low voltage caused by distribution of power over a long distance.
4. High or low voltage supplied by the utility company.

Most *appliance* transformers are capable of adjusting voltage only; they cannot change the frequency of a supply circuit. However, in most instances, 60-Hz equipment can be operated from a 50-Hz supply if the voltage is reduced approximately 8 to 10%. For example, 120-V, 60-Hz equipment can usually be operated on 50 Hz at 108 V.

SELECTING THE PROPER TRANSFORMER

The transformer capacity or rating is the amount of VA (VA = volts × amperes) or wattage of continuous load the unit is capable of operating.

The transformer rating in VA (watts) must always be as large or larger than the load which it is operating to avoid overloading. Overloading causes excessive heating, reducing the life of a transformer.

Small electric motors, portable electric heaters, home appliances, refrigeration equipment, and other electrical equipment carry the volt-

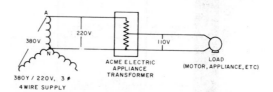

FIGURE 12-9 Line supply voltage that does not match the appliance motor nameplate rating may be corrected by using a transformer as shown here.

age rating on their nameplates. The minimum VA size of the transformer required to operate this equipment is obtained by multiplying the nameplate volts times the amperes. Example: 120 V × 2 A = 240 VA.

To make the proper transformer selection, the load must be completely analyzed, and every electrically energized component must be included. Ten percent should be added to compensate for the high starting current requirement of ac motors and similar items and for inadvertent overloading.

In the interest of economy, it is recommended that where several appliances or adjacent pieces of equipment of the same voltage rating are to be operated, the loads be combined and that one larger transformer be selected. This is less expensive than purchasing several smaller units for individual installation.

STEPS IN MAKING SELECTION

1. Determine the value of the incoming line supply voltage and frequency (50 or 60 Hz).

2. Obtain the appliance or load equipment voltage rating and amperes from the nameplate or instruction sheet accompanying the appliance. Multiply the two to obtain the VA requirement of the load. If the power requirements are listed only in watts, consider this the same as VA. However, on electric discharge lighting and if only wattage is shown, double this figure to obtain the VA rating of the transformer to use.

3. Add all VA requirements of the equipment to obtain the total load. Remember that all components must be of the same voltage rating.

4. Add 10% for high starting current and overloading to obtain the VA size of the transformer.

5. Select a transformer from the charts supplied by the manufacturer using a combination of the supply voltage, the voltage rating of the equipment, the load VA rating, and the type of connection desired.

An example of a manufacturer's data sheet or chart is shown in Fig. 12-10.

Group D

Electrical Tap Connections.
Use on supply consistently above or below nominal voltage.

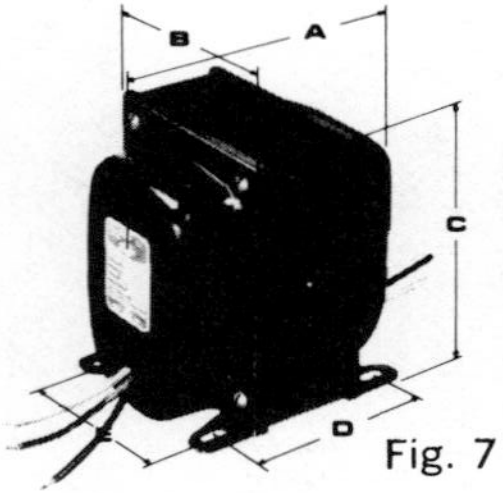

Fig. 7

PRIMARY VOLTS

Tap Conn.	50 Hz	60 Hz
High	230-260	253-283
Med.	210-239	230-260
Low	189-217	208-236

NOMINAL SECONDARY VOLTS

50 Hz	60 Hz
105	115

CATALOG NO.	VA RATING	OUTPUT AMPS	A	B	C	D	E	FIG.	SHIP. WT. #	CONNECTIONS
T-60818	350	3.04	5	$3\frac{5}{8}$	$3\frac{15}{16}$	3	$2\frac{1}{2}$	7	7	10" leads, including ground
T-60820	550	4.78	$5\frac{1}{2}$	$3\frac{11}{16}$	$4\frac{11}{16}$	$3\frac{3}{8}$	3	7	$11\frac{1}{4}$	" "
T-60822	750	6.51	$6\frac{15}{16}$	$3\frac{15}{16}$	$4\frac{11}{16}$	$4\frac{1}{2}$	3	7	$16\frac{3}{4}$	" "
T-60824	1000	8.7	$7\frac{1}{4}$	$4\frac{5}{8}$	$5\frac{5}{8}$	$4\frac{7}{8}$	$3\frac{1}{2}$	7	$23\frac{1}{4}$	" "
T-60826	1500	13.04	$8\frac{7}{16}$	$4\frac{1}{2}$	6	$5\frac{11}{16}$	$3\frac{5}{8}$	7	$28\frac{3}{4}$	" "

FIGURE 12-10 Manufacturer's data sheet for typical instrument transformers.

13

Transformer Installation, Care, and Operation

In this chapter we shall cover general recommendations for the operation and maintenance of dry-type distribution and power transformers of the ventilated and nonventilated type and also liquid-filled distribution and power transformers.

The successful operation of these transformers is dependent on proper installation, loading, and maintenance as well as on proper design and manufacture. As with all electrical apparatus, neglect of certain fundamental requirements may lead to serious trouble, if not to the loss of the equipment. Transformers require less care and attention than almost any other kind of electrical power apparatus, but they still should not be neglected. The condition under which they operate will determine, to some extent, the frequency with which they should be inspected. A regular program of inspection should be established and rigidly carried out.

HANDLING

The safest way to handle large enclosed transformers is to lift from the provisions on the core clamps. It is necessary to remove all or part of the top cover. With the cover removed, it is possible to lower a chain or cable into the enclosure and attach to the lifting devices on the top core clamp. The bottom core clamp is bolted to the base of the enclosure, and the enclosure will be lifted along with the core and coils.

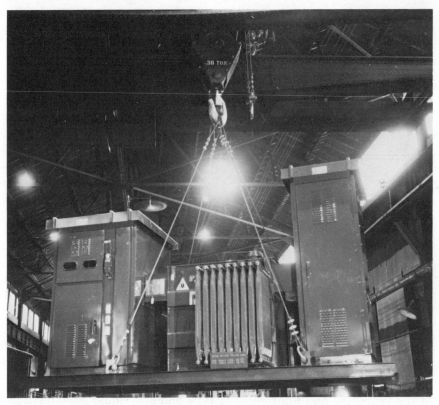

FIGURE 13-1   Oil-cooled transformer being hoisted into position.

If it is undesirable to remove the top cover due to enclosure construction or location, the unit can be lifted from the base frames. The lifting cables must be held apart by a spreader to avoid damaging the enclosure.

Units consisting of core and coil have lifting devices provided only on the core clamp. Special care should be taken when moving the core and coil units as the base is usually rather narrow in comparison to the height of the core and coils.

It is suggested that the weight of the transformer be noted before lifting. This weight is stamped on the nameplate. Do not attempt lifting the transformer by hooking to a bus bar, cable leads, or similar parts. Lift only by means of the lifting devices. Do not allow the lifting cables or chains to press against the bus bars or other parts that are mounted above the core. When lifting, apply tension gradually to the chains, cable, or rope. Do not jerk, jar, or attempt to move the transformer suddenly.

When a transformer cannot be lifted by a crane, it may be skidded or moved on rollers, but care must be taken not to damage the base or tip it over. When rollers are used under large transformers, skids must be used to distribute the stress over the base.

Jacking lugs, two on each long side of the enclosure base, are provided for raising larger transformers with their enclosures. The transformer should be jacked evenly on all four corners to prevent tipping over. Skids or rollers may be placed under the base and the moving accomplished in that manner.

Care must be taken to prevent nails, bolts, nuts, washers, tools, etc., from dropping into the coils. Any metal or foreign matter lodged in the coil ducts is certain to cause insulation failure.

If it is necessary to handle ventilated dry-type transformers outdoors during inclement weather, they should be thoroughly protected against the entrance of rain or snow.

STORAGE

Preferably, ventilated dry-type transformers should be stored in a warm, dry location with uniform temperature. Ventilating openings should be covered to keep out dust. If a unit is stored at a construction site or in a heavily dust-laden atmosphere, it may even be necessary to seal up ungasketed portions of the enclosure. If it is necessary to leave a transformer outdoors, it should be thoroughly protected to prevent moisture and foreign material from entering. Condensation and absorption of moisture can be prevented or greatly reduced by the installation of space heaters or other small electric heaters which maintain the enclosure at a slightly higher temperature than the surrounding ambient temperature.

INSTALLATION

LOCATION Factors which should be kept clearly in mind in locating dry-type transformers are personnel safety, accessibility, ventilation, atmospheric conditions, and locations affecting sound level.

In planning the installation, a location should be selected that will comply with all local and national safety codes and that will not interfere with the normal movement of workers, trucks, or other equipment and material. The location should not expose the transformer to possible damages from cranes, trucks, or moving equipment. It should be remembered that a dent in the enclosure may reduce the insulation clearances to an unsafe level.

As an added safety precaution, thought should be given to the possibility of workers inserting rods, wire, etc., through the ventilation openings of the enclosure and thus coming in contact with live parts. Most transformer ventilation openings are designed in accordance with NEMA standards which require that a $\frac{1}{2}$-in.-diameter rod cannot be inserted through ventilation openings. If this is not sufficiently restrictive, additional requirements should be noted at the time the unit is ordered. Standard enclosures are designed for a maximum average temperature rise of 65°C. In a 35°C ambient temperature, enclosure surfaces may reach 100°C. Precautions should be taken to avoid accidental contact.

The installation will be simplified if any outline drawing is requested. By studying the overall, mounting, and terminal dimensions, it is possible to plan the installation of an orderly arrangement with connections made most conveniently.

Core and coil units (without a case) usually have mounting and terminal dimensions to suit the customer's enclosure. That enclosure should have protection from the coils and have adequate clearances and sufficient ventilation openings.

Ventilated dry-type transformers normally are designed for installation indoors in dry locations. They will operate successfully where the humidity is high, but under this condition it may be necessary to take precautions to keep them dry if they are shut down for appreciable periods. This is discussed more fully later in this chapter. Locations where there is dripping water should be avoided. If this is not possible, suitable protection should be provided to prevent water from entering the transformer case. Precautions should be taken to guard against accidental entrance of water, such as might be obtained from an open window, by a break in a water or steam line, or from use of water near the transformers.

Adequate ventilation is essential for the proper cooling of these transformers. Clean, dry air is desirable. Filtered air may reduce maintenance if the location presents a particular problem. When transformers are installed in vaults or other restricted spaces, sufficient ventilation should be provided to hold the air temperature within established limits when measured near the transformer inlets. This usually will require approximately 100 ft^3 of air per minute per kilowatt of transformer loss. The area of ventilation openings required depends on the height of the vault, the location of openings, and the maximum loads to be carried by the transformers. For self-cooled transformers, the required effective area should be at least 1 ft^2 each of inlet and outlet per 100 kVA of rated transformer capacity, after deduction of the area occupied by screens, gratings, or louvers.

Dry-type transformers should be installed in locations free from unusual dust or chemical fumes. Avoid attaching objects of any kind directly to the case of totally enclosed nonventilated units. Reducing the clear surface area of these units limits cooling capacity and may cause an overheating condition. Transformers should be located at least 12 in. away from walls and other obstructions that might prevent free circulation of air through and around each unit unless the unit is designed for wall mounting and installed per factory recommendations. The distance between adjacent transformers should be not less than this value. Also, accessibility for maintenance should be taken into account in locating a transformer. If the transformer is to be located near combustible materials, the minimum separations established by the National Electrical Code should be maintained.

The transformer case is designed to prevent the entrance of most small animals and foreign objects. However, in some locations, it may be necessary to give consideration to additional protection.

SOUND LEVELS   If noise is a factor in the location and operation of any transformer, special consideration should be given to the installation of the equipment. Many locations can result in an amplification of the sound level. Rooms with low ceilings, closets, and corner locations are examples of areas which may increase the sound level of a transformer.

If a unit is installed in a quiet hallway, you may notice a definite hum. If the unit is installed in a location it shares with other equipment such as motors, pumps, or compressors, the transformer hum will probably go unnoticed.

The transformer is designed to produce a minimum sound level when the following directions are followed:

1.   Connections to primary and secondary terminals made with flexible connectors

2.   All transit bolts and shipping braces removed so the unit will float on rubber isolation pads

3.   All enclosure hardware tightened so panels do not vibrate

INSPECTION   When the transformer has been located at its permanent site, a thorough final inspection should be made before any necessary assembly is accomplished and the unit is energized.

Careful examination should be made to be sure that all external electrical connections have been made properly and that the correct ratio exists between low-voltage and high-voltage windings. This can be tested by applying a low voltage (240 or 480 V) to the high-voltage winding and measuring the output at the low-voltage winding.

It should be determined that all control circuits, if any, are operational and will withstand a 1200-V applied insulation test for 1 min (if the transformer has current transformer circuits, they should be closed).

The operation of fans, motors, thermal relays, and other auxiliary devices should be checked. Fan rotation should be visually verified, and any indicator lights should be checked.

GROUNDING The transformer core and coil assembly and case should be permanently and adequately grounded.

Grounding is necessary to remove static charges and as a precaution in case the transformer windings accidentally come in contact with the core or enclosure. Make certain that the flexible grounding jumper between the core and coil assembly and case is intact or that the core and coil assembly is directly grounded from the core clamp through a flexible lead. Ensure that grounding or bonding meets NEC and local codes.

BOLTED CONNECTIONS If the unit is equipped with standard nonplated aluminum terminals, there should be an abrasive, conductive grease on the bolting surface (example: Alcoa #2 joint compound). This grease should not be removed. The grease should be wire-brushed into the bolting surface area before connecting the input or output bus or cable. This will break up any oxidation that has formed on the aluminum. The washers supplied with the unit must be installed with the dish or cupped side on the aluminum terminals.

Copper or plated aluminum terminals do not require grease, but all aluminum terminals must have Belleville washers.

OPERATION

PLACING IN SERVICE After following the preceding instructions, the transformer may be energized. It is recommended that the unit first be energized at no load followed by a stepped or gradual application of load until full loading is reached. If it is not possible to graduate the load, then full load may be applied.

PARALLEL OPERATION When operating transformers in parallel, their rated voltages, impedances, and turn ratios ideally should be the same. Their phasor relationships must be identical. If these parameters are different, circulating current will exist in the circuit loop between these units. The difference in impedances should in no case exceed 10%. The greater the differences in these parameters, the larger the magnitude of the circulating current. When specifying a transformer to be operated in parallel with existing units, all these parameters should be noted.

LOADING The maximum continuous load a transformer can handle is indicated on the nameplate. However, many specially designed units have specific load capabilities designed into them. If there is any question concerning the load capability of the unit, the factory should be consulted.

MAINTENANCE

PERIODIC INSPECTION Like other electrical equipment, all transformers require maintenance from time to time to assure successful operation. Inspection should be made at regular intervals and corrective measures taken when necessary to assure the most satisfactory service from this equipment.

The frequency at which these transformers should be inspected depends on operating conditions. For clean, dry locations, an inspection annually may be sufficient. However, for other locations such as may be encountered where the air is contaminated with dust or chemical fumes, monthly inspections may be required. Usually after the first few inspection periods a definite schedule can be set up based on the existing conditions.

With the transformer de-energized, enclosure panels should be removed. Inspection should be made for dirt, especially accumulations on insulating surfaces or for those which tend to restrict airflow; for loose connections; for the condition of tap changers or terminal boards; and for the general condition of the transformers. Observation should be made for signs of overheating and of voltage creepage over insulating surfaces as evidenced by tracking or carbonization.

Evidence of rusting, corrosion, and deterioration of the paint should be checked and corrective measures taken where necessary.

Fans, motors, and other auxiliary devices should be inspected and serviced during these inspection periods.

JACKSCREW ASSEMBLY ADJUSTMENTS Check each jackscrew assembly for proper torque by attempting to move the coil block with the hand from side to side. If motion exists, tighten the jackscrew assembly following the outlined procedure. (Caution should be observed when handling nuts, bolts, and washers so as not to drop them into the coils. See Fig. 13-2.)

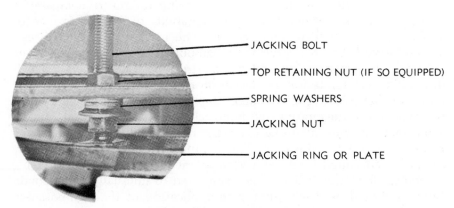

JACKING BOLT

TOP RETAINING NUT (IF SO EQUIPPED)

SPRING WASHERS

JACKING NUT

JACKING RING OR PLATE

FIGURE 13-2 Jackscrew assembly.

1. Tighten the lower jacking nut while holding the jacking bolt until the coil block can no longer be moved by hand.

2. Tighten the lower jacking nut an additional half turn.

3. Apply high-temperature air-dry varnish to the nut and bolt assembly.

4. Repeat as required on other jackscrew assemblies.

CLEANING   If excessive accumulations of dirt are found on the transformer windings or insulators when the transformer is inspected, the dirt should be removed to permit free circulation of air and to guard against the possibility of insulation breakdowns. Particular attention should be given to cleaning the top and bottom ends of the winding assemblies and to cleaning out the ventilating ducts.

The windings may be cleaned with a vacuum cleaner, a blower, or compressed air. The use of a vacuum cleaner is preferred as the first step in cleaning, followed by the use of compressed air or nitrogen. The compressed air or nitrogen should be clean and dry and should be applied at a relatively low pressure (not over 25 psi). Lead supports, tap changers and terminal boards, bushings, and other major insulation surfaces should be brushed or wiped with a dry cloth. The use of liquid cleaners is undesirable because some of them have a solvent or deteriorating effect on most insulating materials.

TESTING   Tests may be made before placing a transformer in service to determine that it is in satisfactory operating condition and to obtain data for future comparison:

1. Insulation resistance

2. Dielectric tests in the field in accordance with 12-02.040 of American Standard C57.12

The insulation resistance test is of value for future comparative purposes and also for determining the suitability of the transformer for application of the high-potential test. The insulation resistance tests should be made before applying the high-potential test. Variable factors affecting the construction and use of dry-type transformers make it difficult to set limits for the insulation resistance. Experience to date indicates that 2 MΩ (1-min reading at approximately 25°C) per 1000 V of nameplate voltage rating, but in no case less than 2MΩ total, may be a satisfactory value of insulation resistance for the application of the high-potential test. However, the manufacturer of the transformers should be consulted for a definite recommendation. If a transformer is known to be wet or if it has been subjected to unusually damp conditions, it should be dried out before application of the high-potential test or before placing it in service, regardless of the insulation resistance.

In addition to the insulation resistance and high-potential dielectric tests, the following tests may be made if desired:

1. Ratio tests for the full windings and all tap positions
2. Resistance measurements of windings
3. Polarity or phase relation
4. Power factor of insulation

If any of these tests are made, it is preferable that they be made before applying the dielectric tests.

As in the case of insulation resistance, insulation power-factor tests may be of value for comparative purposes in checking the condition of a transformer periodically. It should be noted that the power-factor tests on dry-type transformers will be higher than liquid units because of measuring the air as a dielectric.

DETERMINING DRYNESS   The measurement of insulation resistance is of value in determining the status of drying. Measurements should be taken before starting the drying process and at 2-hr intervals during drying. The initial value, if taken at ordinary temperatures, may be high even though the insulation may not be dry. Because insulation resistance varies inversely with temperature, the transformer temperature should be kept approximately constant during the drying period to obtain comparative readings. As the transformer is heated, the presence of moisture will be evident by the rapid drop in resistance measurement. Following this period, the insulation resistance will generally increase gradually until near the end of the drying period when it will increase more rapidly. Sometimes it will rise and fall through a short range before steadying, because moisture in the interior of the insulation is working out through the initially dried portions. A curve, with time as abscissa and resistance as ordinate, should be plotted, and the run should be continued until resistance levels off and remains relatively constant for from 3 to 4 hr.

METHODS OF DRYING   As long as the transformer is energized, humidity conditions are unimportant. In the event that a dry-type transformer is de-energized and allowed to cool to ambient temperature, consideration must be given to the possible effects of humidity.

If the shutdown period occurs during low-humidity conditions, no special precautions should be required before energizing the unit.

Experience indicates that if a shutdown exceeding 24 hr occurs during a period of high humidity, particularly if atmospheric conditions are such as to cause condensation within the housing, then precautions should be taken. Small strip heaters may be placed in the bottom of the unit shortly after shutdown to maintain the temperature of the unit a few degrees above that of the outside air. If such precaution has not been taken, the unit should be inspected for evidence of moisture, and insulation resistance should be checked. If there is evidence of moisture or if the insulation resistance is low, the transformer should be dried out by one of the methods described.

DRYING OF CORE AND COIL ASSEMBLY When it is necessary to dry out a transformer before installation or after an extended shutdown, under relatively high-humidity conditions, one of the following methods may be used:

1. External heat
2. Internal heat
3. External and internal heat

Before applying any of these methods, free moisture should be blown or wiped off the windings to reduce the time of the drying period.

DRYING BY EXTERNAL HEAT External heat may be applied to the transformer by one of the following methods:

1. By directing heated air into the bottom air inlets of the transformer case

2. By placing the core and coil assembly in a nonflammable box with openings at the top and bottom through which heated air can be circulated

3. By placing the core and coil assembly in a suitably ventilated oven

It is important that most of the heated air passes through the winding ducts and not around the sides.

Good ventilation is essential in order that condensation will not take place in the transformer itself or inside the case. A sufficient quantity of air should be used to ensure approximately equal inlet and outlet temperatures.

When using either of the first two external heating methods, heat may be obtained by the use of resistance grids or space heaters. They may either be located inside the case or box or may be placed outside and the heat blown into the bottom of the case or box. The core and coil assembly should be carefully protected against direct radiation from the heaters.

It is recommended that the air temperature be at least 100°C but should not exceed 150°C.

DRYING BY INTERNAL HEAT This method is relatively slow and should not be used if one of the other two methods is available.

The transformer should be located to allow free circulation of air through the coils from the bottom to the top of the case. One winding should be short-circuited, and sufficient voltage at normal frequency should be applied to the other winding to circulate approximately normal current.

It is recommended that the winding temperature not be allowed to exceed 100°C, as measured by resistance or by thermometers placed in the ducts between the windings. The thermometers used should be of the spirit type; mercury thermometers give erroneous readings due to the generation of heat in the mercury resulting from induced eddy currents. The end terminals of the windings (and not the taps) must be used in order to circulate current through the entire winding. Proper precaution should be taken to protect the operator from dangerous voltage.

DRYING BY EXTERNAL AND INTERNAL HEAT This is a combination of the two methods previously described and is by far the quickest method. The transformer core and coil assembly should be placed in a nonflammable box, or kept in its own case if suitable, and external heat applied as described in the first method and current circulated through the windings as described in the second method. The current required will be considerably less than when no external heating is used but should be sufficient to produce the desired temperature of the windings. It is recommended that the temperatures attained not exceed those stated in the foregoing.

DETERMINING DRYING TIME Drying time depends on the conditions of the transformer, its size, its voltage, the amount of moisture absorbed, and the method of drying used as explained previously.

LIQUID-IMMERSED TRANSFORMERS

The following paragraphs cover general recommendations for the operation and maintenance of liquid-filled distribution and power transformers. While this information specifically covers liquid-immersed transformers manufactured by Hevi-Duty, most of the recommendations are characteristic to all liquid-filled distribution and power transformers.

INSTALLATION

LOCATION Installation location of a transformer must be considered carefully. Transformers, as is the case with most electrical equipment, generate a substantial amount of heat during operation. This heat must be removed in order to allow the transformer to maintain its designed maximum temperature limits. If a transformer is located outdoors, the heat will be removed by natural convection cooling unless the radiator airflow is restricted by surrounding objects.

Indoor installations require adequate ventilation to remove the heat of transformer operation. Inlet ventilation openings should be as low as possible and outlet ventilation openings as high as possible. Care should be taken to provide an average ambient temperature around the transformer of 30°C unless the transformer is specifically designed for higher ambients. Care should also be taken to prevent restriction of air circulation. Adequate space must be maintained between transformers or between a transformer and nearby equipment or walls. Separation is especially important near the transformer radiators, with a spacing equal to the radiator panel depth being recommended.

ASSEMBLY Transformers, with equipment or accessories removed for shipment, must be reassembled after being placed on the installation site. They should be reassembled in the following order.

DETACHABLE RADIATORS Inspect all radiator panels and flange mating surfaces for shipping damage.

Check that all valves on tank flanges are closed, and remove blank shipping plates, using care not to damage the gasket. (Note the tank flange match mark numbers.)

Remove blank shipping plates on radiator flanges, and inspect for moisture or contamination inside radiator headers. If the radiators are contaminated, flushing will be necessary.

Clean all mating surfaces on the tank and radiator flanges. If gaskets are not in place on the tank flanges, apply a small amount of rubber cement to hold them in place during installation of the radiators.

Lift the radiators by means of the single lifting eye at the top. Install the radiators with matching numbers on tank flanges. Bolts should be drawn up evenly, alternating across corners, top, and bottom, until spring washers are compressed. Tighten each nut a half turn further.

Flush radiators if they are contaminated. Do not open the tank flange valves prior to flushing the radiators. Remove the top and bottom pipe plugs from the radiator headers using a filter press. Reverse the flushing procedures so that the radiators are flushed top to bottom and then bottom to top. Reinstall the pipe plugs after flushing using Teflon thread sealing tape.

Relieve the tank pressure or vacuum, and vent the tank by removing a hand hole cover, shipping plate, or plug, whichever is most convenient (the oil level may be above normal for shipment, so this vent opening should be on the top cover). Open first the bottom and then the top flange valve on each radiator in succession until all valves are open. After all radiators are installed, the unit should be reevacuated and topped off to the proper (25°C) cold oil level.

BUSHINGS   Remove the blank bushing plates, using care not to damage the gasket. Draw leads, if used, will be attached to the underside of the blank plate.

Secure the lead connector for the draw lead bushings with a length of wire at least 12 in. longer than the bushing. Remove the bushing top cap hardware, and insert the wire up through the bushing. Pull the draw lead connector into place while lowering the bushing onto the flange. Install the locking pin at the top of the bushing, and remove the pull wire. Install and tighten the bushing flange hardware to apply even pressure to the flange gasket. Install and tighten the gasket and the top cap assembly.

With fixed bushing studs and connectors, transformer leads inside the tank must be connected after the bushing is secured to the flange. If necessary, the oil level must be lowered to provide entry into the tank.

PRESSURE-VACUUM GAUGE   Remove the pipe plug, usually located on the tank front and about 5 in. below the top cover. Install the gauge and tighten using Teflon sealing tape.

FANS   Attach the fans to the radiators using T connectors between the panels. The fans will be located, generally, at the top of the radiator panels.

RAPID PRESSURE RISE RELAY   Remove the $2\frac{1}{2}$-in. pipe plug on the tank top or side, and install the rapid pressure rise relay. This relay should be positioned with the lead connector down for proper operation. Connect the flexible lead connector between the control box and the rapid pressure rise relay connector.

LIGHTNING ARRESTORS   Lightning arrestors and lightning arrestor brackets will be mounted in accordance with the manufacturer's recommendations. Care should be taken that all ground connections are securely made in accordance with all applicable local and national codes.

CLOSING AND FILLING   A final internal inspection should be made on any transformer before it is energized, particularly if any work has been done inside the tank. All electrical connections should be checked for tightness. All the bushings should be checked for tightness of the gaskets, and all draw lead connections should be checked. All electrical clearances inside the tank should be checked. One final check should be made that all tools, or any extra materials that have been used inside the transformer, have been removed.

Reinstall all hand hole covers which have been removed. All gasket grooves should be cleaned, with all gaskets in the correct position.

All nuts on the bushing flanges and hand hole covers should be torqued to the following values:

$\frac{3}{8}$-in. nuts: 20 ft-lb

$\frac{1}{2}$-in. nuts: 50 ft-lb

If the oil level has been lowered for inspection or if the unit was shipped without being completely filled with oil, the unit must be filled to the proper level before energization. The proper liquid level will be noted on the nameplate.

On tanks designed for full vacuum processing, a vacuum may be taken during filling or prior to final purging. All accessories that may be damaged by vacuum should be removed and the remaining openings covered with solid covers or plugs. These accessories include the pressure-vacuum bleeder and the pressure relief device. A vacuum can then be obtained through the top vacuum connection.

TESTING FOR LEAKS The simplest method for testing for leaks is by gas pressure. The gas space in the unit should be pressurized at 5 psi with dry nitrogen. The gas pressure should be monitored for a period of approximately 24 hr. A change in pressure does not necessarily indicate a leak. Any temperature increase or decrease in the transformer will result in a subsequent increase or decrease of the gas pressure in the unit.

Ambient temperatures and tank pressure should be monitored for the 24-hr period.

If there is a significant drop in pressure during the 24-hr period without an accompanying decrease in ambient temperature, the tank must be checked for leaks. Repressurize the tank at 5 psi. Using a solution of liquid soap and soft water, brush all weld joints above the oil level, all bushing gasket flanges, and all hand hole cover gaskets. Any leaks in the gas space above the liquid will be shown in the gas form of soap bubbles. Use chalk dust below the liquid level to check for leaks of liquid from the tank.

DETERMINING DRYNESS   The core and coils of all transformers are thoroughly dry when they are shipped from the factory, and every precaution is taken to ensure that dryness is maintained during shipment. However, due to rough handling or other causes, moisture may enter the transformer and be absorbed by the oil and insulation. It should, therefore, be determined that the oil and insulation are dry before the transformer is energized.

If the transformer has been shipped with the core and coils immersed in oil, samples of the oil should be drawn from the bottom sampling valve and tested for dielectric strength. If the oil tests at 26 kV or more and there is no evidence of free water in the bottom of the transformer, it can be assumed that the insulation is dry, and the transformer can be energized.

A transformer shipped in nitrogen should be tested for oxygen content before it is unsealed. If gas pressure is present and the oxygen content is less than 5%, it can be assumed that the core and coils are dry, and the transformer can be energized.

If the tests indicated low dielectric strength or high oxygen content, further investigation should be made to determine the cause before the transformer is energized.

If there is no conclusive means of measuring the dryness of the transformer in the field, it is recommended that the insulation power factor and resistance measurements be taken and submitted to the factory for recommendations. To obtain a uniform insulation temperature, the transformer oil should be at normal ambient temperature when the insulation power-factor measurement is made. The top and bottom oil dielectric test results should accompany the power-factor reading.

If the tests or visual inspection indicate the presence of moisture, the core and coils must be dried before voltage is applied to the transformer.

FINAL EXTERNAL INSPECTION All external surfaces of the transformer, and accessories, should be examined for damages that may have occurred during shipment or handling. The liquid level gauge, thermometer, pressure-vacuum gauge, tap changer, and other accessories should be checked for proper operation. Bushings should be checked for cleanliness and, if necessary, should be cleaned with xylene or other nonresidual solvent.

All valves should be checked for proper operation and position. Radiator valves, if supplied, should be in the open position. If a conservator tank is supplied, the connection between this tank and the main tank should be open. The upper filter press valve should be closed.

All liquid levels should be checked, including those in any oil-filled switches or conservator tanks, if supplied. The conservator tank should also be properly vented. All electrical connections to the bushings should be checked for tightness. Proper external electrical clearances should be checked. All cables or buses connected to the transformer bushings should be checked to avoid strain on the porcelain insulator. All winding neutral terminals should be checked to assure that they are properly grounded or ungrounded, according to the system operation. All tank grounds should be checked. All current transformer secondaries should be checked to assure that they are either loaded or short-circuited.

The tap changer should be padlocked in the correct position for operation. All cooling fans and control circuits should be checked for proper operation.

Obtain a sample of liquid and check it for dielectric strength. The liquid should be filtered if it tests low.

OPERATION

PLACING INTO SERVICE After applying full voltage, the transformer should be kept under observation during the first few hours of operation under load. After several days, check the oil for oxygen content and dielectric strength.

All temperatures and pressures should be checked in the transformer tank during the first week of operation under load.

PARALLEL OPERATION If transformers are to be used in parallel, it is important to check the nameplates to make sure that they are suitable for parallel operation. The following characteristics must be checked for parallel operation:

1. Voltage ratios must be within .5%.
2. Vector relationships must be identical.
3. Impedance should be the same within plus or minus $7\frac{1}{2}\%$.

Current should be carefully monitored between both units to make sure that one unit is not carrying all the load under parallel operation. The units should be monitored for an additional period of at least 1 week to make sure that there is no abnormal temperature rise on either unit.

LOADING Except for special designs, transformers may be operated at their rated kVA if the ambient temperature of the cooling air does not exceed 40°C for any 24-hr period and the altitude does not exceed 3300 ft.

MAINTENANCE

PERIODIC INSPECTION The following is a checklist of the more important points to be checked at least annually and preferably semiannually:

1. Determine that the oil level in the transformer tank and all liquid-filled compartments, such as junction boxes or switches, is

satisfactory. Test the dielectric strength of the liquid. Oil from the tank bottom that tests 24 kV or less should be filtered.

2. Measure the oxygen content in transformers with inert gas systems. If the oxygen content is 5% or more, the transformer should be purged with dry nitrogen.

3. Check the nitrogen bottle content on positive pressure inert-gas systems.

4. Clean all bushings if dirty, and inspect the porcelain for cracks. Check the oil level on oil-filled bushings.

5. Check the pressure relief device, if furnished.

6. Check the thermometer, liquid level gauges, pressure gauges, and other indicators.

7. Check the thermometer drag pointer to see if there is evidence of excessive loading at some time in the past.

8. Make a megger check or power-factor check of insulation and bushings for comparison with previous observations.

9. Clean the fan blades, and check fan operation by turning the control switch to "manual."

10. Check the paint on the tank and accessories, and repaint when required.

11. Make certain that no tools or other objects have been left in, or on, the transformer.

12. Close all openings after completion of inspection. Purge with clean, dry nitrogen, and repressurize to 3 psi.

INSULATION POWER-FACTOR AND RESISTANCE MEASURE-MENTS   Regular insulation resistance or power-factor measurements provide a means of observing and recording changes in the insulation due to moisture accumulation or chemical deterioration. Insulation resistance and power-factor measurements are also necessary in indicating the progress of drying a transformer or its oil.

Every measurement should be taken carefully, using the same procedure in each case to lessen chances for inconsistency.

Do not use an instrument having a voltage output in excess of the voltage rating of the winding being tested. Record the readings every 2 hr when measurements are made in connection with drying a transformer. When vacuum drying, take readings after each vacuum period and before and after filling with oil.

Before taking measurements, make sure the bushings are clean and dry, as dirty porcelain may cause low readings.

POWER FACTOR   Short-circuit all windings at the bushing terminals when measuring the power factor. All windings except the one being tested should be thoroughly grounded. No windings should be allowed to "float" during the measurement. Any winding which is solidly grounded must have the ground removed before the power factor can be measured on that winding. If this is not possible, do not include the winding in the power-factor measurements, as it must be considered part of the ground circuit.

Power-factor readings should be taken for each winding to all other windings and ground.

Examples of the readings to record for a three-winding transformer are the following:

1. HV: LV, TV, GND
2. LV: HV, TV, GND
3. TV: HV, LV, GND

Temperature affects power-factor readings considerably. Therefore, it is necessary to determine the insulation temperature at the time the readings are taken for correct interpretation. It is usually sufficient to take top and bottom oil temperatures. When checking the top oil, use a spirit thermometer rather than a mercury one, as there is less danger to the transformer in case of breakage. The bottom oil temperature can be measured by placing a thermometer in a stream of oil drained from the bottom filter valve.

If the power factor is checked without oil in the transformer, measure the temperature by the increase in conductor resistance method. Otherwise, obtain readings from thermometers or thermocouples placed at various locations on the windings and insulation. Remember, the main concern is the temperature of the insulation, not the conductor. Examine each situation closely, and devise some method of determining the insulation temperature with reasonable accuracy.

INSULATION RESISTANCE Insulation resistance can be measured with a megger or megohm bridge. Be sure the scale of the instrument reads higher than the insulation resistance being measured.

Insulation resistance measurements, taken without oil in the transformer, have greater significance in determining dryness. When measured with oil in the transformer, insulation resistance is considerably affected by the condition of the oil, since oil resistance is measured in parallel with solid insulation. In general, insulation resistance with oil in the transformer is much lower. The insulation resistance changes by a large amount with variations in temperature. It may be high when measured at room temperature, even though the insulation is not dry.

Insulation resistance measurements will vary widely from transformer to transformer. An approximate minimum value for insulation resistance is 25 MΩ/kV of rated line-to-line voltage.

During the drying process, insulation resistance measurements are necessary and should be taken at 2-hr intervals at fairly constant temperatures. Both the resistance and temperature of the insulation should be recorded. Short-circuit and ground all windings except the one being tested.

When meggering, take the reading after the megger voltage has been applied to the winding for about a minute. Keep this period of time consistent for all readings throughout the drying process.

INTERPRETATION OF MEASUREMENTS Power-factor measurements are the most reliable in determining dryness and should be taken in preference to insulation resistance, especially in large and high-voltage transformers.

As the drying proceeds at a constant temperature, the insulation power factor will generally decrease. Finally it will level off and become reasonably constant when the transformer becomes dry. In some cases, the power factor may rise for a short period early in the drying process. The insulation resistance will generally increase gradually until near the end of the drying process; then the increase will become more rapid. Sometimes the resistance will rise and fall through a short range one or more times before reaching a steady high point. This is caused by moisture in the interior parts of the insulation working through the portions that have already dried.

The drying process should be continued for approximately 12 hr after the insulation power factor becomes consistently low and the insulation resistance becomes consistently high.

When vacuum drying is used, it may be more difficult to obtain insulation power-factor and resistance measurements at convenient temperatures. Such irregularities, however, do not outweigh the value of drying the transformer by this method.

It is recommended that in case of questionable readings, the log of insulation power-factor and resistance readings with time and temperatures be submitted to the factory for comments. Include in this information the transformer serial number, a description of the measuring instruments used, the drying out procedure, the methods of taking temperature readings, and any other pertinent data.

METHODS OF DRYING There are a number of approved methods of drying out transformer core and coils, any of which will be satisfactory if carefully performed. But remember, if the drying out process is carelessly or improperly performed, great damage may result to the transformer insulation through overheating.

DRYING BY CIRCULATING CURRENT Circulating a current through the windings drives the moisture from the core and coils into the oil. This moisture is in turn removed from the oil by filtering or evaporation. This is the slowest of the approved methods.

Before attempting the drying process, any free water should be drained from the transformer through the drain valve at the base of the tank. Water that is in suspension with the transformer oil should be removed by means of a filter press. Discontinue the use of the filter press during the drying process if the oil temperature is too high for efficient filter press operation. The efficiency of the filter press is greatest at 20°C to 40°C.

Heat the windings and oil to a high temperature by short-circuiting the low-voltage windings, and apply an ac voltage to the high-voltage windings. Apply voltage sufficient to produce one-half to three-quarters of full rated load current in the windings.

The voltage necessary to produce the desired short-circuit current can be determined from the following formula:

$$ESC = \% \, IZ \times E \text{ rated} \times .01$$

where

ESC = required voltage to produce full rated load current

$\% \, IZ$ = percent impedance volts (from nameplate)

E = maximum rated voltage of winding to which power is applied

The oil flow should be restricted to raise the temperature. If the transformer is equipped with radiator valves, one of these (preferably the top) should be closed on each radiator. If the transformer does not have radiator valves, it should be blanketed. Otherwise, drop the oil level below the upper radiator opening. Make sure that all high-voltage parts are under oil.

Table 13-1 indicates the short-circuit current in percent of full rated load current which may be applied for this method of drying transformers with the corresponding maximum allowable top oil temperature in degrees centigrade. A regulator should be connected in series with the winding to which the voltage is applied to provide current control.

The top oil temperature must not exceed the values shown for the given loads. The winding temperature is higher than that of the oil, and damage to the insulation may result if the values given in Table

TABLE 13-1 Short-Circuit Current in Percent of Full Rated Load Current

Self-cooled Rating	Max. Permissible Top Oil Temp.
60%	85°C
75%	80°C
100%	75°C

13-2 are exceeded. Check the top oil temperature carefully at regular intervals.

The tank cover and top part of the tank, to the oil level, should be thoroughly blanketed outside in order to prevent moisture condensation inside the tank. The amount of blanketing required below the oil level must be determined by trial.

Good ventilation should be maintained at the top of the tank so water vapor will dissipate and not condense on the underside of the cover and on other surfaces inside the tank. This can be accomplished by raising or removing the hand hole or manhole cover and protecting the openings from the water. The cover should still be blanketed with insulating material to protect against moisture condensation.

WHEN TO DISCONTINUE DRYING   During the drying process, samples of the insulating liquid should be taken every 4 hr, from both the bottom and top of the tank, and tested for dielectric strength. The samples should be allowed to cool to 40°C or below before the test is conducted. The drying should be continued until the liquid from the top and bottom of the tank tests at least 26 kV for oil for seven consecutive tests, with the insulating liquid maintained at maximum temperature for the given load.

A decrease in the dielectric strength indicates that moisture is still passing from the core and coils into the oil. The drying process should be continued until the tests show constant or increasing dielectric strength.

After the drying process is completed, close all openings in the transformer cover and add clean, dry oil, if needed, to bring the oil level up to normal. This additional oil should be added at the top of the transformer to prevent the formation of air bubbles which might become trapped in the windings if the oil is added through the drain valve.

The transformer should now be operated at approximately two-thirds voltage for 24 hr with no load. If at the end of 24 hr the dielectric strength of the oil is satisfactory, the transformer may be put on the line.

TABLE 13-2   Characteristics of Insulation Oil

Gravity	27.4 API
Flash	154°C
Color	L0.5
Pour	-57°C
Viscosity:	59.3 Saybolt
38°C	59.3 Saybolt
Carbon residue	.01
Neut. no.	.02
Ash	.003
Dielectric kV min.	30

Insulation power factors and resistance readings should be recorded during the drying process, along with the time and temperature at which the readings are made. Drying a transformer with oil in the tank is a very slow and tedious method. Use only for small units or when other methods are impossible.

DRYING BY EXTERNAL HEAT The movement of clean, dry, heated air through the core and coils is the method employed in this drying process.

Remove the core and coils from the tank and place in a box with holes in the top and near the bottom to allow proper air circulation. The clearance between the sides of the transformer and the box should be small so the heated air is forced to pass up through the ventilating ducts in the coils and not around the sides. Apply heat at the bottom of the box.

It may be possible to dry the transformer in its own tank with the oil removed by using radiator and drain valve openings to attach piping from a blower and an external heat source. Baffles may be needed between the core and coils and the tank walls to force the heated air up through the ventilating ducts.

The best way to obtain heat is from resistance heaters. The temperature of the air going into the transformer should never exceed 100°C. The core and coils must be carefully protected against direct radiation from the heaters. It is also advisable to completely line the box, when used, with asbestos. When forced air is used, suitable baffles should be placed between the heater and the transformer enclosure inlet.

Do not circulate current in the windings when air drying, as the time gained is not great enough to justify the fire hazard. Great care must be taken to ensure that no point on the core and coils exceeds 100°C. Place several spirit-type thermometers or thermocouples among the coils, preferably in the tops of the ventilating ducts, and screen them for protection against air currents.

As the temperature rises quite rapidly at first, readings should be made about every $\frac{1}{2}$ hr. To keep the transformer at a constant temperature for insulation resistance and power-factor measurements, one thermometer should be in a position where it can be read without being moved. The other thermometers should be moved around until the hottest spots are found. Then keep them at these points throughout the drying period. Wherever possible, check the insulation temperature by the increase in resistance method.

Adequate extinguisher equipment of the inert-gas type, such as carbon dioxide or nitrogen, should always be on hand when drying out a transformer.

WHEN TO DISCONTINUE DRYING   Power-factor and insulation resistance readings, along with the time and temperature at which the readings are made, should be recorded throughout the drying process. The drying process should continue until the power-factor and insulation resistance readings are essentially constant with a constant temperature for a 12-hr period. When the drying process has been completed, the core and coils should be immediately immersed in clean, dry oil.

DRYING WITH HEAT AND VACUUM   In this method of drying, the core and coils are heated to between 85°C and 90°C and then a vacuum is drawn to reduce the boiling point of water and to remove the vapor formed. This method assures quicker and more thorough and even drying and should be used whenever possible.

Using the procedure as outlined previously, heat the core and coils by circulating current through the windings with the transformer filled with oil. Hold the top oil temperature between 75°C and 85°C for about 2 hr; then de-energize the transformer and drain the oil as rapidly as possible. Seal the transformer and immediately draw a vacuum to obtain full advantage of the heat. Hold the vacuum for approximately 4 hr or until the winding temperature drops to between 50°C and 60°C.

Insulation power-factor, resistance, and temperature readings should now be taken and the transformer refilled with dry filtered oil under vacuum. After filling with oil, check the insulation power-factor and resistance measurements. Repeat this procedure until the measurements show the insulation to be dry.

The number of times that this procedure must be completed will depend on the condition of the transformer, the temperature of the insulation when the tank is being evacuated, and the completeness of the vacuum. Normally, three to seven cycles will be necessary.

As complete a vacuum as possible should be drawn, but it should never exceed the maximum for which the tank is designed.

Considerable drying time can be saved by keeping the oil hot while it is out of the tank. Short-circuit heating will not be required if the transformer can be filled with oil at a maximum temperature of 85°C. The oil should remain in the transformer for about 2 hr and then be removed and a vacuum drawn. The maximum allowable temperature for heating oil outside the tank is 100°C. If the oil is heated inside the transformer by circulating current in the windings, the maximum temperature is 85°C.

Where it is not possible to transfer the oil back and forth between the transformer and storage, an alternate method of vacuum drying may be possible. The core and coils can be heated in the tank without oil by circulating heated air through the drain valves, as described in the section on drying by external heat. Heat should be applied until the winding temperature is between 80°C and 85°C by resistance measurement. The transformer should then be sealed and a vacuum drawn and held for approximately 4 hr.

Before the vacuum is released, take insulation power-factor, resistance, and temperature readings. Either oil or dry nitrogen should be used to fill the evacuated tank. Ambient air should not be allowed to enter the transformer to break the vacuum. The rapid cooling of air entering the vacuum may cause condensation of moisture inside the transformer. Reapply heat and repeat the cycle until the readings indicate that the transformer is dry. Refill immediately with clean, dry transformer oil under vacuum.

CARE OF OIL

Table 13-2 shows the characteristics of insulation oil.

HANDLING AND STORAGE OF TRANSFORMER OIL Because the sulfur in a natural rubber hose dissolves in oil, causing the dielectric strength to be lowered, metal or oilproof hoses or pipes must be used for handling transformer oil. Dissolved sulfur also deteriorates the conductor in transformer windings.

Containers of oil should be stored in a closed room having a constant temperature. If stored outside, they must be protected from the weather. Drums should be placed on their sides with their bungs down at a 45° angle and tightly closed.

Unless tests are required, drums or other containers should not be opened until the oil is to be used. Before opening, be sure that the oil temperature is as high or higher than that of the surrounding air to prevent condensation or moisture. Containers that are to be filled with transformer oil should be thoroughly cleaned and rinsed with the liquid before they are used.

TESTING The dielectric strength of liquid should always be checked before putting it into the transformer. After filling the transformer, samples should be taken for a dielectric strength test.

SAMPLING OF TRANSFORMER OIL A large-mouth glass bottle with a cork or glass stopper should be used for collecting samples of transformer oil. Before using the bottle, clean it with xylene or other nonresidual solvent and dry it well. Rinse the container several times with the oil to be tested before collecting the sample. If a dielectric test only is to be made, 1 pint of transformer liquid will be sufficient; however, if other tests are to be made, drain off 1 quart.

University of Glamorgan
Learning Resources Centre -
Treforest
Self Issue Receipt (TR1)

**Customer name: MR TSHEPISO
MOSIKARI**
Customer ID: **********901

Title: Mastering electronic and
electrical calculations
ID: 7311328665
Due: 17/09/2010 23:59

Title: Handbook of power
generation : transformers and
ID: 0022979703
Due: 17/09/2010 23:59

Total items: 2
18/05/2010 14:29

Thank you for using the Self-
Service system
Diolch yn fawr

Test samples should not be taken until the oil has settled. This time varies from 8 hr for a barrel to several days for a large transformer. Cold oil settles more slowly and not as completely as warm oil. Always take samples from the sampling valve at the bottom of the tank or storage drum.

When sampling, drain off sufficient liquid to be sure that a true specimen is obtained and not one that may have collected in the pipes. A clear glass container is best for observing the presence of free water and other contaminants. If any are found, an investigation should be conducted to determine the cause and the situation remedied.

Although water may not be present in sufficient quantity to settle out, a considerable amount of moisture may be suspended in the oil. The oil should therefore be tested for dielectric strength. Care must be taken to prevent contaminating the oil sample after it has been collected. The sample should be taken on a clear, dry day when the oil is as warm or warmer than the surrounding air. A small amount of moisture from condensation or other causes may produce a poor test.

TESTING DIELECTRIC STRENGTH   A standard cup for liquid testing should be used when checking the dielectric strength. Clean the cup thoroughly and rinse with a portion of the liquid to be tested. The liquid and the gap receptacle should be normal room temperature, or about 25°C. Tip the sample container and swirl the liquid a few times before filling the test cup to aid in mixing impurities which might be present in the sample. Avoid vigorous agitation which might introduce an excessive amount of air into the liquid. Completely fill the test cup and allow 3 min for air to escape before applying voltage.

Voltage should be increased at a rate of about 3000 V/sec. One breakdown should be made on each five fillings of the test cup. Any individual test which deviates from the average by more than 25% should be disregarded and replaced by an additional test. The average of the first five tests within the allowable deviation can be considered to represent the dielectric strength of the liquid.

If the dielectric strength of the sample tests below 22kV for oil, collect a new one, making certain that the liquid does not become contaminated after it is collected.

The minimum dielectric strength of the oil is 26 kV when it is shipped. Oil that tests below this should not be put into the transformer. Oil that tests below 24 kV in service should be filtered or reprocessed.

FILTERING   When filling a transformer with liquid, filtering is recommended to prevent dirt, lint, and moisture from entering the tank.

A filter press is effective for removing all types of foreign matter, including finely divided carbon and small deposits of moisture. Begin the filtering process with new blotter paper and replace it frequently, depending on the amount of moisture removed. Blotter paper must be thoroughly dried and kept warm until the time it is used.

Lose no time when transferring filter paper from the oven to the press. Hours of drying time can be wasted if the filter paper is exposed to the air more than a few minutes. Common practice used in changing blotter paper is to

1. Remove one sheet from the inlet side of each set.
2. Replace with a clean piece on each outlet side.
3. Repeat the procedure about every $\frac{1}{2}$ hr.

Keep a close check on the dielectric strength of the filtered liquid. If tests indicate moisture in suspension, all the blotters must be removed before filtering can continue.

To extract free water and sizable amounts of moisture, a centrifuge is more practical than a filter press. However, when used in combination (the liquid passing through the centrifuge first), much better results will be obtained for liquid in poor condition.

When liquid is filtered in the transformer or tanks, it is preferable to draw the liquid from one tank, filter it, and discharge it into a clean, dry tank. However, at times it is satisfactory to draw liquid from the bottom filter press valve and, after filtering, return it through the top one. Care must be taken so that no moisture is returned to the top of the tank with the liquid. Aeration of the liquid also must be avoided. Filtering should be done with the transformer de-energized.

If tests show the presence of a large quantity of moisture and dirt, filter the bottom oil separately, drawing it from the transformer into a separate tank. When water and dirt have been removed so that the oil tests 27 kV or greater, change the filter connection to the upper filter press valve and return the liquid to the top of the transformer. Continue filtering from the bottom and returning to the top until the tests reach the accepted standard.

FILLING WITH LIQUID Check the dielectric strength of the liquid while it is still in containers. If free water is present, drain off the water before putting the liquid through the filter press. Continue passing the liquid through the filter press until a dielectric strength of 27 kV or higher for oil is obtained.

NONVACUUM FILLING In cases where vacuum filling is not required, the tank should be filled through the upper filter press connection. A second opening at the top should be provided to relieve the air being displaced. Full voltage should not be applied to the transformer for a period of 48 hr after filling.

VACUUM FILLING Entrapped air is a potential source of trouble in all transformers. In general, therefore, it is desirable to fill transformers with liquid under as high a vacuum as conditions permit. It is essential to vacuum-fill high-voltage transformers shipped in nitrogen in order to develop their full insulation strength before they are energized.

The transformer tank must be airtight except for the vacuum and oil connections. After obtaining a vacuum as high as the tank construction will permit (see the nameplate), this vacuum should be maintained by continuous pumping for at least 4 hr. The filling may then begin. The liquid line should be connected to the upper filter press connection or other suitable connection on top of the tank. The filtered liquid is admitted through the connection with the rate of flow being regulated by a valve at the tank so that the vacuum does not fall below 90% of the original value. Any air bubbles in the liquid will explode in the vacuum, and the air will be drawn out by the vacuum pump. The vacuum should be maintained for 3 or 4 hr after the transformer is full. For small units, the vacuum will not determine the rate of flow.

TROUBLESHOOTING DRY-TYPE TRANSFORMERS

Transformer failures may occur in either the electric, magnetic, or dielectric circuit.

Symptom	Cause
	Electric Circuit
Overheating	Continuous overload; wrong external connections; poor ventilation; high surrounding air temperature (Rating is based on 30°C average temperature over 24-hr period with peaks not to exceed 40°C.)
Reduced to zero voltage	Short turns; loose connections to transformer terminal board
Excess secondary voltage	Input voltage high; dirt accumulations on primary terminal board
High conductor loss	Overload; terminal boards not on identical tap positions
Coil distortion	Coils short-circuited
Insulation failure	Continuous overloads; dirt accumulations on coils; mechanical damage in handling; lightning surge
Breakers or fuses opening	Short circuit; overload
Excessive cable heating	Improper bolted connection

Symptom	Cause
High voltage to ground (using rectifier or VTVM meter)	Usually a static charge condition
	Magnetic Circuit
Vibration and noise	Low frequency; high input voltage; core clamps loosened in shipment or handling; loose hardware on enclosure; shipping braces and/or hold-down bolts not removed; transformer location
Overheating	High input voltage
High exciting current	Low frequency; high input voltage; shorted turns
High core loss	Low frequency; high input voltage
Insulation failure	Very high core temperature due to high input voltage or low frequency
	Dielectric Circuit
Smoke	Insulation failure
Burned insulation	Lightning surge; switching or line disturbance; broken bushings, taps, or arrestors; excess dirt or dust on coils
Overheating	Clogged air ducts or inadequate ventilation
Breakers or fuse open	Insulation failure

If any of the preceding symptoms are noticed, the transformer should be immediately removed from service. Immediate attention may save a large repair bill. Many times the trouble can be quickly determined and the transformer returned to service.

If the trouble cannot be definitely corrected, no further use should be made of the transformer until the cause has been found.

It may be necessary to remove the core and coils for a closer examination. If no apparent fault can be found, the core and coils may have to be taken apart for a detailed inspection. Removal of the coils from the core is usually a factory or service shop operation.

TROUBLESHOOTING LIQUID-IMMERSED TRANSFORMERS

Transformer failures may occur in either the electric, magnetic, or dielectric circuit.

Symptom	Cause
	Electric Circuit
Overheating	Continuous overload; wrong external connections; poor ventilation; high surrounding air temperature (Rating is based on 30°C average temperature over 24-hr period with peaks not to exceed 40°C.)
Reduced or zero voltage	Shorted turns; loose internal connections; faulty tap changer
Excess secondary voltage	Input voltage high; faulty tap changer
Coil distortion	Coils short-circuited
Insulation failure	Continuous overloads; mechanical damage in handling; lightning surge
Breakers or fuses opening	Short circuit; overload; inrush current, internal or external
Excessive bushing heating	Improper bolted connection
High voltage to ground (using rectifier or VTVM meter)	Usually a static charge condition
	Magnetic Circuit
Vibration and noise	Low frequency; high input voltage; core clamps loosened in shipment or handling
Overheating	High input voltage
High exciting current	Low frequency; high input voltage; shorted turns
High core loss	Low frequency; high input voltage
Insulation failure	Very high core temperature due to high input voltage or low frequency
	Dielectric Circuit
Pressure relief Device operation	Insulation failure
Burned insulation	Lightning surge; switching or line disturbance; broken bushings, taps, or arrestors
Overheating	Inadequate ventilation
Breakers or fuse open	Insulation failure
Bushing flashover	Environmental contaminants; abnormal voltage surge
	Mechanical
Cracked bushing	Overstress due to cable load; mechanical handling
Loss of pressure	Check gaskets, cracked bushing, welds

If any of the preceding symptoms are noticed, the transformer should be immediately removed from service. Immediate attention may save a large repair bill. Many times the trouble can be quickly determined and the transformer returned to service.

If the trouble cannot be definitely corrected, the transformer should be taken out of service until the cause has been found.

It may be necessary to remove the hand hole cover for a closer examination. If no apparent fault can be found, the core and coils may have to be removed for a detailed inspection. Removal of the core and coils is usually a factory or service shop operation.

14

Pole
and Platform Mounting
of Transformers

The most popular transformers for pole and platform mounting at the present time are oil-immersed, self-cooled distribution transformers. They are used at every point where the high-voltage distribution line needs to be stepped down to the required secondary or service voltage.

In any particular voltage class, the actual voltage of a transformer has increased in recent years. For example, the 2400-V class of transformers formerly were rated 2200 = 110/220, then later they were rated 2300 = 115/230, and today they are rated 2400 = 120/240 V. This gradual increase in the rated voltage of transformers also occurred in the other voltage classes.

In the early days of urban electrical distribution, practically all systems were 2400-V-class delta systems, and the 2400-V transformer was designed and manufactured for this system. The selection of 2400 V for distribution was logical from the standpoint of service and economy, as this voltage is high enough to give good system performance on systems where the distribution is very long. In addition, the voltage is sufficiently low to result in economical distribution equipment.

In recent years most 2400-V delta systems have been changed over to 2400/4160Y-V systems. This change was due to the fact that as the 2400-V delta systems became more heavily loaded, it became necessary to put in larger distribution line conductors or raise the operating voltage to maintain proper voltage regulation. The most economical procedure in this case was to raise the operating voltage to 4160Y, and this was economical because the change did not necessitate a change in transformers or other equipment on the line.

POLE MOUNTING OF TRANSFORMERS

It has become common practice to mount single transformers up to and including 15 kVA directly on the pole itself with pole mounting brackets; otherwise they are mounted with the use of crossarms and crossarm hanger irons. Individual and banks of transformers up to and including $37\frac{1}{2}$ kVA and individual 50-kVA transformers can be crossarm mounted. Banks of 50 to 100 kVA and in some instances smaller ones are mounted on H-frame-type platforms.

Crossarm mounting structures will vary, but one type consists of a set of double arms placed at a required distance below the primary circuit arm, from which the transformers are suspended, with a single or kick arm placed at an appropriate distance below the double arms to support the lower end of the hanger irons.

Porcelain fused cutouts or hookstick-operated fused disconnecting switches are used to accomplish transformer overcurrent protection and disconnecting means. They are usually mounted on the back arm of the double-arm structure or on a separate single arm. In the lower distribution voltages, the cutout is sometimes mounted on the line arm, particularly in the case of single-transformer installations.

When lightning arrestors are required, they are normally mounted as follows:

1. On the transformer hanger crossarm, usually on the opposite side of the single arm or set of double arms from the used cutout
2. Directly on the transformer by means of a metal bracket when the transformer is mounted directly on the pole or free standing on a substation base
3. On the line or buck arm when a pole riser from an underground section of the system is involved

The primary tap conductors are installed from the primary line conductor to the cutout and transformer. They are supported on pin insulators when it becomes necessary to maintain the proper clearances. Secondary circuits are installed at levels either below or above the transformer on crossarms or secondary racks. The secondary tap conductors are installed from the transformer secondary connection terminals directly to the secondary line conductors without insulator supports.

When the H-frame transformer platform structures are used, the primary buses are provided by suspending the three conductors horizontally with suspension insulators between crossarms above the transformers. The fused cutouts and lightning arrestors are placed on one of the bus crossarms, and primary taps are installed from the primary line conductors to the cutouts and then on to the bus conductors. Taps are installed from the buses to the primary terminals on the transformer. The secondary buses are usually provided in the same manner as for the primary buses except that larger conductors of proper size are used and the buses are installed either below or to one side of the transformers. All bus conductors are usually installed with insulated conductors.

Some utility companies use transformers with primary wiping sleeve entrances and single-conductor lead-covered cables terminated at the line end in potheads. They are installed on the cutout double-arm structure and connected on the transformer end with lead-wiped connections. Such installations reduce the chances of an accident by the workers on the platform when they come in contact with hot primary conductors.

Distribution transformers are manufactured with and without high-voltage taps to adjust for line voltage drop and with and without an internal overcurrent protective device. The type of mounting for other items of equipment mounted on pole structures depends on the type and size of equipment.

POLE RISERS

There are many instances such as the transmission and distribution systems and providing customer services when it is necessary to connect an underground section of the system to an overhead section. This is handled by some type of *pole riser.*

The following are the basic requirements of a pole riser:

1. To protect the underground cable on the pole from physical injury
2. To provide a means of terminating the cable and the conductors and protecting the insulation from entrance of moisture
3. To provide a means of disconnecting the underground section of the system from the overhead section and providing lightning protection when necessary
4. To protect linepersons working on the pole from accidental contact with grounded portions of the pole riser such as the conduit or cable sheath

When the pole riser extends from an underground raceway system, it is normally extended from a manhole or pull box near the riser pole. When the pole riser extends from a direct burial underground system, the cable is carried to the pole base and enters the riser raceway through an elbow. When the pole riser is to provide a customer service, it extends from the main switchboard or customer substation location to and up the pole.

It is necessary to protect the cable from physical injury, and this is accomplished either by extending the necessary size of conduit the full length of the riser from the ground level to the appropriate crossarm level in which the cable is installed or by extending the conduit up the pole to approximately 10 ft above grade and then enclosing the conductors from that point to the level of the appropriate crossarm with a wooden molding or box.

The following are the means of terminating the cable conductors and protecting the insulation from the entrance of moisture:

1. In the case of high-voltage distribution and transmission systems, when cables with a lead sheath and insulation susceptible to moisture are used, cable terminators or potheads are used. Where large sizes of cables of this type are used for low-voltage distribution or customer service, cable terminators or potheads may also be required. Potheads may be of the single-conductor or multiconductor types depending on the type of cable. When multiconductor potheads are used, they may be attached directly to the riser raceway by means of a conduit fitting and clamping ring and stuffing box to secure the cable or cables, or the raceway may be terminated a short distance below the pothead and the cable sheath joined to the pothead by means of a wiping sleeve entrance. When single-conductor potheads are used, the individual cables are "fanned out" from the raceway and attached to the potheads through wiping sleeve entrances. Jumper connections are made from the pothead terminal cap nuts to the disconnecting or other equipment or to the overhead system conductors by means of short lengths of wire.

2. Low-voltage lead–sheathed cables and rubber or synthetic jacketed cables in general are properly taped and insulating paint applied at the point from which the insulation is removed from the conductor to prevent the entrance of moisture. The conductors are connected directly to the disconnecting or other equipment or to the overhead system conductors.

3. When rubber-and-braid or synthetic covered building wire and cable are used, a standard service entrance head may be required to terminate the riser raceway at the crossarm level.

Disconnecting the underground section of the system from the overhead section may be accomplished by using individual hookstick-operated disconnect switches, or porcelain-enclosed fused cutouts, or gang-operated pole top disconnecting switches.

Linepersons working on the pole must be protected from accidental contact with the grounded riser raceway or cable sheath. This is accomplished by placing over the raceway a section of wooden molding or box from the top of the raceway to a required distance below the lowest crossarm or by the wooden molding or box placed over the conductors when that means is used to protect them from physical injury.

Pole riser raceway and cable sheaths must be adequately grounded.

STEEL TOWER INSTALLATIONS

Steel tower construction is the type most often used for high-voltage transmission lines over long distances and over various types of terrain. The primary purpose of steel tower construction is for the support of conductors and not the mounting of equipment. The transformers and switching equipment are normally located at substations at the generating plants or at intervals along the way.

Steel towers provide stronger supporting structures than wood poles, allow greater height of structures, and are not subject to damage by fire. They are essentially self-supporting and do not require guys and anchors as in the case of wood pole structures.

Steel towers are erected on concrete footings or foundations. The size and type of footing depend on the type and height of the tower and the nature of the ground and terrain. Footings are usually completely detailed in the contract working documents.

In general, steel tower footings require the following operations:

1. Clearing or grading the tower base area
2. Locating the individual footings
3. Excavating
4. Building concrete forms when required
5. Installing reinforcing rods
6. Pouring the concrete
7. Removing the forms and backfilling around the footings

The tower itself is erected by assembling individual pieces of galvanized steel of various sizes and shapes; they are bolted together. The pieces have been fabricated at the mill in accordance with drawing details for the tower and shipped to a central job location.

At the central location, the steel is sorted and identified. Usually all the pieces of the steel for a tower are transported to the tower site, where the tower is constructed in place by bolting together the component parts progressively from bottom to top using wood booms and gin poles or self-propelled cranes. Some sections of the tower may be subassembled on the ground and hoisted into position by a crane.

Suspension-type insulators are used almost entirely in steel tower construction.

SUBSTATIONS

Substations primarily serve the purpose of transforming the voltage of the system and providing overcurrent protection, lightning protection, disconnecting means, and cross-switching to provide interconnection or circuits.

Overcurrent protection is usually accomplished by means of oil circuit breakers, although in some instances of lower circuit voltages and capacity, special fuses are used. Disconnecting means is provided by gang-operated, high-voltage air-break switches or individual hookstick-operated disconnect switches for lower voltages. Lightning protection is provided by various types of lightning arrester equipment. Cross-switching is provided by gang-operated, double-throw air-break switches.

15

Saturable Core
Reactors

A saturable core reactor is a magnetic device having a laminated iron core and ac coils similar in construction to a conventional transformer and is uniquely effective in the control of all types of high power-factor loads. The coils in a saturable core reactor are called gate windings. In addition it is designed with an independent winding by which direct current is introduced; this is the control winding.

When the ac coils of a saturable reactor are carrying current to the load, an ac flux (magnetism) saturates the iron core. With only ac coils functioning, the magnetism going into the iron core restricts the ac flow to the load. This restriction of current flow (impedance) causes the voltage output to the load to be about 10% of the line supply voltage. Since the iron core is always fully saturated with magnetic flux, the use of direct current from the control winding introduces dc flux, which displaces the ac flux from the iron core. This action reduces the impedance and causes the voltage output to the load to increase. By adjusting the dc flux saturation, the impedance of the ac gate windings may be infinitely varied. This provides a smooth control ranging from approximately 10 to 94% of the line voltage at the load. See Fig. 15-1.

Since there is practically no power loss in the control of the impedance in a saturable reactor, a relatively small amount of direct current can control large amounts of alternating current.

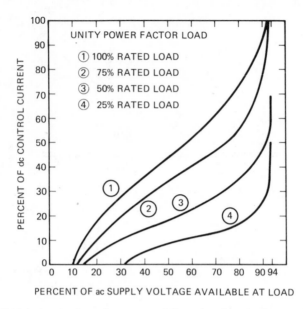

FIGURE 15-1 Load voltage (in percent of line voltage) versus dc control in percent of rated current).

APPLICATIONS

Saturable core reactors eliminate the need for mechanical and resistance controls and are a very efficient means of proportional power control for resistance heating devices, vacuum furnaces, infrared ovens, process heaters, and other current-limiting applications.

In lighting control—especially where wattage per circuit is large—a saturable reactor eliminates the loss of dissipated power of a resistance control and provides an infinitely smooth regulated power output to the lighting load.

Wound rotor motors can be started smoothly and operate at speeds commensurate with the load when a saturable reactor, connected in series with the motor, provides the control. This, of course, completely eliminates the maintenance of costly grid resistors and drum controllers.

TYPES OF SATURABLE REACTORS

Although details of design will vary with each manufacturer, usually three basic styles are available. Small-size reactors are constructed with two ac coils (gate windings) on a common core with the dc control winding on the center leg of the core, as shown in Fig. 15-2.

Figure 15-3 shows a medium-size reactor which is constructed

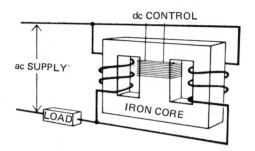

FIGURE 15-2 Small-size reactors are constructed with two ac coils on a common core with the dc control winding on the center leg of the core. (Courtesy Acme Electric Corporation.)

with two ac coils each mounted on an independent wound core. The dc control winding is independent but encompasses both coils.

The larger-size reactors normally utilize four ac coils as the gate windings, as shown in Fig. 15-4. In addition, the dc control coil constitutes a continuous winding which circumscribes each pair of ac coils.

Each pair of ac coils is connected in parallel in a manner to prevent the dc windings from being affected by induced ac voltage, harmonics, or high peak voltages when sudden load change occurs.

The matched design of the gate winding and dc control winding results in a highly efficient magnetic coupling. The ac output of a saturable reactor with matched windings, connected to a unity power-factor load, can achieve in some designs as high as 97% of the input line voltage.

Saturable reactors constructed with wound cores, in which the lamination end joints are equally distributed over a wide portion of the

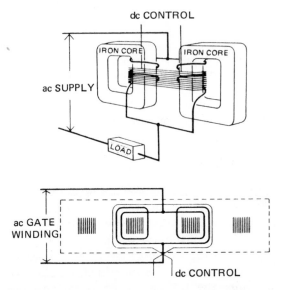

FIGURE 15-3 Medium-size reactors are constructed with two coils each mounted on an independent wound core. The dc control winding is independent but encompasses both coils. (Courtesy Acme Electric Corporation.)

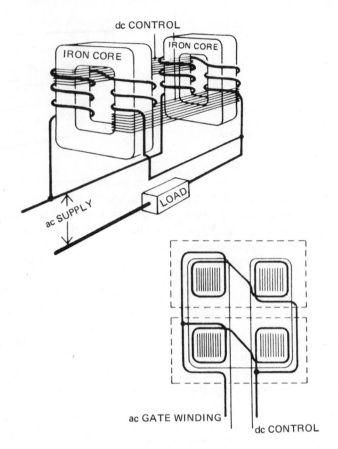

FIGURE 15-4 Larger-size reactors utilize four ac coils as the gate windings. In addition, the dc control constitutes a continuous winding which circumscribes each pair of ac coils. (Courtesy Acme Electric Corporation.)

core's circumference, provide an unusually wide voltage adjustment for the load, particularly in the lower-voltage range.

RANGE OF CONTROL

A saturable core reactor connected in series between the line supply and the load will provide a maximum line voltage to the load in relation to the power factor of the load. However, in no case will the available load voltage ever be 100% of the line voltage.

Since inductance is the principal electrical property of a saturable reactor, the current-impedance drop added to the normal current-resistance drop limits the percentage of voltage that will be transferred through the reactor. Because of this normal voltage loss, the electrical units comprising the load, based on their power factor, should be suitable to operate at voltages less than line voltage.

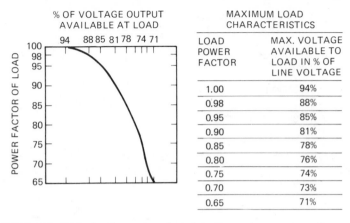

% OF VOLTAGE OUTPUT AVAILABLE AT LOAD	MAXIMUM LOAD CHARACTERISTICS	
	LOAD POWER FACTOR	MAX. VOLTAGE AVAILABLE TO LOAD IN % OF LINE VOLTAGE
	1.00	94%
	0.98	88%
	0.95	85%
	0.90	81%
	0.85	78%
	0.80	76%
	0.75	74%
	0.70	73%
	0.65	71%

FIGURE 15-5 With the dc supply adjusted to its maximum power, the maximum voltage to the load will be slightly less than the line supply voltage, depending on the power factor and the percent of rated load.

With the saturable reactor connected to the load and without dc excitation, the normal exciting current in the gate winding will supply a voltage to the load of approximately 10% of the line voltage value. As the dc control is increased in power, the ac voltage to the load is increased. With the dc supply adjusted to its maximum power, the maximum voltage to the load will be slightly less than the line supply voltage, depending on the power factor and the percent of rated load. See Fig. 15-5.

This relationship between applied dc control and ac output can best be illustrated in the wiring diagrams in Figs. 15-6, 15-7, and 15-8.

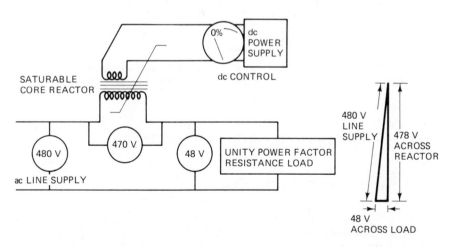

FIGURE 15-6 With no direct current applied, voltage across the reactor reads approximately 98% of line voltage, and the voltage at the load is approximately 10% of line voltage.

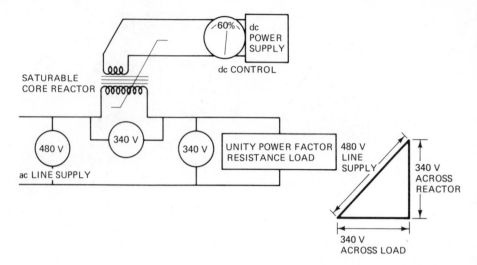

FIGURE 15-7 With 60% direct current applied, voltage across the reactor measures 70% of line voltage, and the voltage at the load also reads 70% of line voltage.

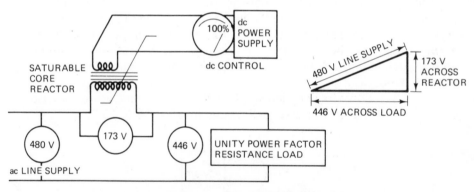

FIGURE 15-8 When 100% direct current is applied, the voltage across the reactor drops to 37% of line voltage, and the voltage to the load increases to 94% of line voltage.

The triangles graphically illustrate the relationship of the ac voltage values of the supply line, across the reactor, and at the load.

DETERMINING REQUIREMENTS

Saturable reactors are rated in kVA equal to the unity power-factor load they control. To determine the reactor size required for a given load, use the following formula for single-phase applications:

$$\text{saturable reactor kVA} = \frac{94\% \text{ of line voltage} \times \text{ load current}}{1000}$$

To illustrate the use of this equation, assume a voltage of 240 V with a load current of 12 A single phase. Substituting these values in the equation, we have

$$\frac{94\% \times 240 \times 12}{1000} = 2.71 \text{ kVA}$$

A 3-kVA saturable core reactor would be satisfactory for the preceding application provided the load is unity power factor. However, if the load is less than unity power factor, the reactor kVA required should be calculated in accordance with the power-factor constants that follow:

Load Power Factor	Power-Factor Constant Multiplier
1.00	1.00
.98	1.09
.95	1.17
.90	1.29
.85	1.42
.80	1.55
.75	1.70
.70	1.84
.65	1.89

Using this table, the load kW $\times$ power-factor constant (in the table) = kVA size of the saturable reactor. For example, if the power factor is 80% (using the previous example of load), the load of 2.71 kVA $\times$ 1.55 power-factor constant = 4.2 kVA—the adjusted size needed for this application. For this application, use a 5-kVA saturable reactor, which is the nearest standard larger size.

SINGLE-PHASE LOAD CONTROL

Saturable reactors can be used to control individual circuits of a lighting installation, even to control individual spotlights, or a single reactor can be used to uniformly control the light output of a multiple-circuit installation, as shown in Fig. 15-9.

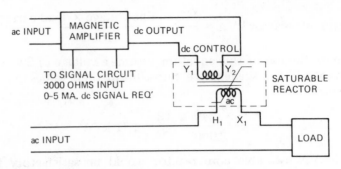

FIGURE 15-9 Diagram showing connections of a single-phase saturable reactor and magnetic amplifier controlling a single-phase load.

THREE-PHASE APPLICATIONS

Three single-phase reactors, each in series with each supply line, can be used to control a three-phase load, as three-phase reactors are not commercially feasible. Since each reactor controls a third of the total power to the load, each is rated at one-third of the total load. In other words, three 2-kVA single-phase reactors will control a 6-kVA three-phase load, as shown in Fig. 15-10.

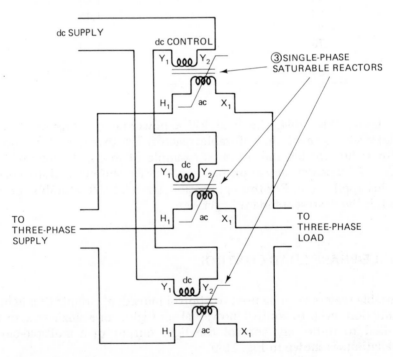

FIGURE 15-10 Typical connection diagram of three single-phase reactors connected to a three-phase load.

Care must be exercised to make certain the proper reactor voltage is used for three-phase loads. For example, for 240-V, three-phase service, use three 138-V single-phase saturable reactors. For 480-V, three-phase service, use three 277-V, single-phase saturable reactors.

The dc control winding of the three reactors should be connected in series to provide symmetrical control of each phase. When this connection is used, the dc control voltage rating of each reactor is one-third of the dc supply voltage. The dc power supply used for controlling the reactors must be capable of delivering three times the wattage required to control one single reactor. Other applications are shown in Figs. 15-11, 15-12, and 15-13.

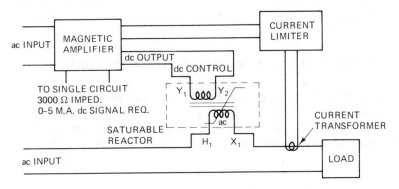

FIGURE 15-11 Diagram showing connections for single-phase current limiting.

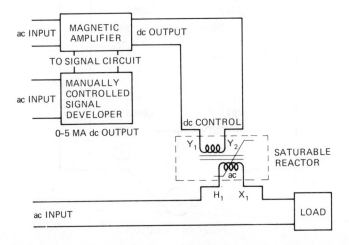

FIGURE 15-12 Diagram showing connections for adding a manually controlled signal developer power supply to supply a signal to the magnetic amplifier.

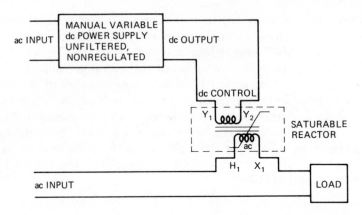

FIGURE 15-13 Diagram showing connections for an unfiltered, manually regulated power supply used in place of the magnetic amplifier to control the saturable reactor.

REVERSING ROTATION OF THREE-PHASE MOTORS

Saturable reactors may be used to reverse the rotation of three-phase induction motors. For this application, they should be connected as shown in Fig. 15-14. Note that four saturable reactors are used but that only two reactors are saturated at any one time. Reactors 2 and 3 pro-

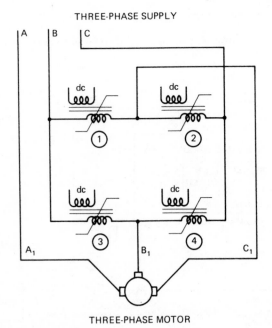

FIGURE 15-14 Diagram showing connections of four saturable reactors to reverse three-phase induction motor.

vide clockwise rotation. When these two are de-energized of direct current and reactors 1 and 4 are saturated, the motor shaft will reverse. This may be accomplished slowly or almost instantaneously. The latter could be used with some modification in the form of a motor brake.

DIRECT-CURRENT REQUIREMENTS

The amount of dc power required for manual or automatic adjustment of the ac load is very small—usually between .33 and 1% of the load. This dc power requirement may be expressed as gain. To illustrate, the gain of a 10-kVA reactor controlled by 60 W of direct current would be the following: 10,000 W/60 W = gain of 167. The gain of a 3-kVA reactor controlled by 30 W of direct current would be the following: 3000 W/30 W = gain of 100. The gain of a 25-kVA reactor controlled by 115 W of direct current would be the following: 25,000 W/115 = gain of 217.

16

Pole-Line Construction

Overhead electric distribution facilities include such work as wood pole overhead line construction, steel tower overhead line construction, substation and switchyard construction, and overhead railway systems.

The use of wood as a means of supporting overhead electrical distribution systems is an economical and satisfactory method of overhead construction for practically all types of transmission lines, street and highway lighting, and most communication systems. The basic components consist of wood poles, pole holes, auxiliary supports, crossarms, crossarm braces, multipole structure cross braces, insulators and related hardware, anchors and guys, assorted hardware and armor rods, transformers and switching equipment, equipment platforms, and pole top structures.

The height or length of poles depends primarily on their use and the terrain on which they are used. The voltage is closely related to the height of the pole, as are the number of circuits, type and extent of equipment mounted on the poles, changes in direction of the line, length of conductor span, and the like. However, the usual lengths of wood poles will range from 20 to 60 ft on most installations. The smaller sizes will be used for secondary distribution of outside wiring for residential, farm, and similar uses. The longer pole sizes will be used for transmission lines and certain applications of distribution lines. Exceptionally long poles may be required for river or canyon crossings, recreation floodlighting, and some special antenna uses.

The most common type of line construction encountered for

carrying secondary distribution lines will be the single-pole structures. This type of system consists of overhead electrical or communication lines using pin-type insulators, insulated clevices, lightweight groups of suspension insulators, or insulator racks attached to individual poles.

The distance between poles is governed by

1. Class and height of the pole
2. Size and strength of conductors
3. Differences in grade elevation of adjacent poles
4. Change in direction of the line and operating conditions
5. Terrain in which the poles are used

For use on building properties where the pole location is governed by the property line and the buildings to be fed, wood poles are seldom spaced more than 150 to 250 ft apart. On the other hand, power company poles used to serve rural areas are often spaced from 300 to 600 ft apart, and in the case of cross-country transmission lines, the spacing is usually greater, that is, from 500 to 800 ft where the conductors are of sufficient size and tensile strength that their weight will not cause them to break under severe weather conditions.

POLE FOUNDATIONS

Wood poles are held in an upright position by means of holes of sufficient depth that are dug or drilled in the ground. The depth of the holes depends on the height of the poles and degrees of change in direction of the line as well as the nature of the soil in which the holes are dug. Obviously, poles set in soft ground will have to be set deeper than those set in hard rocky ground. For poles 25 ft and under, 3-ft holes are common. However, for poles over 25 ft, the minimum depth in earth is not usually less than 5 ft.

POLE SETTING

Line construction crews almost always use an A-frame boom or other modern apparatus for setting poles. However, the plant electrician or electrical contractors who primarily engage in interior wiring usually set poles by hand. This is done with pike poles and a sufficient number of workers for the size of the pole being set.

Poles are usually set so that alternate crossarm gains face in opposite directions, except at terminals and dead ends where the gains of the last two poles should be on the side facing the terminal or dead end. On unusually long spans the poles are usually set so that the crossarm comes on the side of the pole away from the long span. Where single-pole-top pins are used, they are usually on the opposite side of the pole from the gain, with the flat side against the pole. End poles and those set at a change of direction are usually set at a slight angle away from the pull of the conductors, while the poles in between are set plumb or vertical.

At dead ends and changes in direction of the pole line, some measure must be taken to prevent the poles from breaking off and to prevent upsetting, swaying, or misaligning the pole structure. This is accomplished by guys which consist of steel guy cable—one end attached to the pole and the other anchored in the ground or to some other solid structure.

Dead man or log anchors require the digging of a hole large enough to allow the placing of the log, the anchor rod usually being driven through the ground at the angle of the guy, to intercept the anchor log to which it is attached so that the log presses against firm earth when a strain is put on the anchor. In some instances, where poor ground condition exists, a mass of concrete is poured into the hole and around the anchor rod to provide sufficient holding capacity.

Guys are usually attached to the pole structure at or near the level of the conductor crossarm. When more than one circuit is carried on the pole, the guy is either attached to the pole midway between the two conductor crossarms, or an individual guy is attached at each circuit level. Multiple guys can be attached to the same anchor when it has sufficient holding capacity. When circuits are added to an existing pole line and there is doubt as to the holding capacity of the existing anchor, an additional anchor must be installed. An example of this would be the later installation of communication cable on an existing power pole. To prevent persons from accidentally walking into a down guy, a wooden or metal guy guard is placed over or around the guy cable at the ground level.

CONDUCTORS

Individual conductors used for overhead power lines range from solid No. 8 copper to large bare stranded cables. In the larger sizes, various combinations of stranded copper or aluminum and steel cable are used where the copper or aluminum conductor does not have sufficient tensile strength to support its own weight or the added weight of ice in areas where this is a problem. Hard-drawn or medium-hard-drawn conductor is used rather than soft drawn because of greater tensile strength. Weatherproof-covered conductor is used on the lower-voltage distribution secondary circuits and service drops. No. 8 or No. 6 bare solid conductor is usually used on series lighting circuits.

When a given section of conductor has been pulled into place, it is connected at one end, usually to the dead end insulators. The conductor is then given a certain amount of sag between adjacent pole structures. When the proper sagging has been accomplished, the conductor is fastened at the ends and then is connected either to the pin insulators or to the bottom suspension insulator. When the conductor is pulled through pulleys, it is raised to the level of the suspension insulators with block and tackle or *coffing hoists.*

At locations where the line makes a considerable change in direction, the conductor is dead-ended with suspension insulators on each side of the crossarm structure. In such cases, the electrical circuit is made continuous by means of jumpers of short length of conductor either carried over the top of the crossarm structure and supported by pin insulators or suspended under the tower member.

17

Underground Wiring

Underground electrical distribution systems are either buried directly in the earth, or else the conductors are pulled through an underground raceway system.

Direct burial installations will range from the installation of a relatively small single-conductor cable to large multiconductor distribution cables. However, in both cases, the conductors are installed in the ground either by placing them in an excavated trench which is then backfilled or by burying them directly by means of a cable plow which opens a furrow, feeds the conductors into the furrow, and closes the furrow over the conductor.

Where the job site conditions permit, the use of a cable plow is by far the fastest way to install direct burial cable. There are certain conditions, however, that will prevent the use of such a plow; they are as follows:

1. Type of earth in which the cables are to be buried
2. Types of conductors
3. Presence of existing underground facilities
4. Nature of the terrain
5. Contract specifications
6. Amount of work involved and the availability of the equipment

When the conductors are to be installed in a trench, the requirements of the contract should be carefully noted for special installation requirements such as the placing of a layer of sand in the bottom of the trench, removal of stones or rock from the backfill, or the placing of creosoted wooden boards over the conductors for protection during backfilling.

Underground raceway systems consist of manholes or junction boxes and connecting runs of one or more conduits (rigid, PVC, fiber, etc.) placed in trenches and sometimes encased in concrete. The size and number of raceways depend on the use of the system, the number of conductors, and the spare raceway capacity desired but will normally range in conduit size from 2 to 6 in. inclusive.

When an underground raceway system does not require large chambered manholes to serve the purpose of installing and splicing conductors and small items of equipment, relatively shallow junction boxes are often used. They are usually constructed of cast iron, steel, or concrete with a square or oblong steel plate cover.

During the installation, underground raceways are usually installed so that the top tier of conduits is at least 2 ft below the finished grade. From this it is easy to determine the overall depth of the trench by calculating the size of the conduit, the separation between layers, and the distance of the bottom conduit from the bottom of the trench.

SPLICING AND CONNECTING DIRECT BURIAL CONDUCTORS

Direct burial conductors are usually spliced or connected to equipment at equipment bases or in concrete manholes or junction boxes. In such instances, the conductors are carried into the equipment bases, manholes, or junction boxes through short lengths of conduit or ducts often encased in concrete. Where the point of splicing or connection is above the grade level, such as at street and airport lighting and substation locations, elbows must be provided.

Cables are joined in manholes or junction boxes by splices made by a qualified cable splicer. Cables, however, are not always spliced in every manhole as in some instances a run of cable may be pulled through one or more manholes with sufficient slack to allow racking.

The splicing of cables, particularly lead-covered cables and/or the connection of cables to cable entrances of equipment, is a specialized operation which is performed by workers who have acquired the knowledge of and the ability to perform such operations. Through special training and actual experience, the splices and connections must be made in such a manner as to properly insulate the conductors for the system voltage and to prevent the entrance of moisture into the cable insulation or into the equipment as well as to provide a permanently secure electrical and mechanical joining of the conductors or connection to the equipment terminals.

In instances where it is desirable, or where direct burial conductors are required to be spliced for continuing runs or branch runs at points other than in equipment bases, junction boxes, or manholes, special splicing equipment is available for complete direct burial or with covers accessible at the grade level. The types of connection or splicing enclosures will vary with the size and type of cable and the voltage of the system. Markers are usually required to indicate the location of buried cable splices or splice and connection boxes.

When the scope of an underground system or a section of it does not require large chambered manholes to serve the purpose of installing and splicing conductors and small items of equipment, such as series lighting transformers, relatively shallow junction boxes are often used. They are usually constructed of concrete with a square or oblong steel plate cover. In some instances, cable racks and ground rods and drainage provision are required.

TRENCHING METHODS

In general, trenching methods fall into the categories of hand dug, machine dug, or a combination of the two. Digging with a hand pick and shovel is usually limited to short runs, adverse ground conditions, or the existence of other utilities (water pipes, telephone lines, etc.). Otherwise, trenching machines are more often used, as they are more economical than hand digging. Self-propelled mechanical trenching equipment is available to dig the normal range of depth and width of most trenches required for underground wiring systems.

When the location of runs of direct burial conductors requires them to be placed under roadways, airport runways, railroad tracks, and similar obstacles, they are installed in conduit or ducts, usually encased with a concrete envelope both to provide a means of allowing their initial installation and to prevent the rupturing of the conductor insulation and sheath due to the settling of the pavement or the slight depression of the pavement or track under load. When an additional run of direct burial cable crosses over the location of an existing one, a run of conduit of adequate size and length, with fiber bushings placed on each end, may be required to prevent the contact of one cable with another. Raceways placed under existing tracks or paved areas often may be installed with the use of conduit jacking equipment.

Special boring machines are sometimes used to drill or push conduit under roadways or sidewalks. Such machines require only a narrow starting trench and terminal sump hole, making restoration of the area to its original condition less demanding. Drill heads range in size from $1\frac{1}{4}$ to 2 in. for the initial cutting pass and from 2 to $3\frac{1}{2}$ in. for the return or reaming pass.

When drilling under sidewalks or roadways is not practical, normally a suitable channel can be cut with a concrete saw. Two cuts are made with the saw along the path of the conduit or cable. This can be done with little or no damage to the surrounding pavement. The gravel and dirt are then removed to accept the conduit and cable, after which the cut needs only minor patching.

Other special tools include an electronic metal locator for locating underground cable and conduit, an underground fault locator for locating a break or ground fault in buried cable, and a power wire-pulling apparatus for pulling long runs of heavy cable through underground raceways.

UNDERGROUND RACEWAY SYSTEMS

Underground raceway systems consist of manholes or junction boxes and connecting runs of one or more steel conduits or nonmetallic ducts placed in trenches and usually encased with a concrete envelope. The size and number of raceways depend on the use of the system, the number of conductor runs, and the spare raceway capacity desired. Normally, the size of the raceways ranges from 2 to 6 in. inclusive.

Heavy wall nonmetallic (PVC plastic) conduits are available for direct installation in trenches without a concrete envelope. They are usually limited to single conduit runs or multiple conduit runs requiring only one horizontal tier of ducts and where the top area is not used for mounting heavy equipment or material.

Manholes or junction boxes are provided to allow the following:

1. The splicing of continuing conductor runs
2. The facilitation of the installation of continuous conductor runs
3. The junction of lateral runs
4. The installation of operating equipment

When communication and power raceways are included in the same raceway system, two-section manholes are constructed with a dividing wall and separate entrances from the grade level.

While rigid steel conduit may be used in some instances, nonmetallic raceways are normally used for underground systems.

When using conduit encased in concrete, it is customary to drive grade stakes at approximately 25-ft intervals to enable the workers to accurately grade the bottom layer of concrete to the height of the first conduit layer. This concrete is placed and rammed and should be sufficiently dry to bear the weight of the workers, who may begin immediately to lay the bottom tier of conduit.

In using PVC (plastic) conduit, the joints are painted and prepared with cement and the conduit adjusted to the proper line and separation by either wooden comb separators or by plastic. In the case of concrete-encased fiber ducts, concreting to the height of the next succeeding fiber layer is then begun and follows closely the laying of the first fiber tier. As the work progresses, the worker engaged in raking the concrete over the duct removes the wooden comb separators from the concrete and fills the voids. When precast concrete separators, generally $1\frac{1}{2}$ in. square in cross section, are used, they are left in place and covered with concrete. The top of the separator furnishes the grade for the concrete at the height of the next duct layer. A light tamping of the concrete with a flat wooden tamper is advisable to ensure an even surface. If sufficient trench has been excavated, the laying of the fiber is generally accomplished in 100-ft stretches, with the concrete gang working on both sides of the fiber layers, either placing the 3-in. concrete bottom or concreting over a fiber tier previously placed.

Some areas require the setting of the concrete before starting a new fiber tier, while others insist on the placing of a concrete that is comparatively dry and capable of bearing the weight of workers engaged in laying the second and succeeding fiber tiers before the concrete has had time to set. Allowing the concrete to set before laying the next duct tier results in a denser and stronger concrete and straighter alignment but has a disadvantage in that the duct bank consists of a series of layers which are likely to heave and separate under heavy frost conditions. The method of concreting and laying the successive tiers before the concrete has set, when properly performed, has the advantage of obtaining a satisfactory bond between the concrete layers, at the expense of spading and tamping the comparatively dry concrete to ensure a satisfactory dense mass.

The availability of concrete for raceway encasement falls into two general categories:

1. Job site
2. Ready mix

In all but remote areas, ready-mix concrete services are available which, when delivery is properly scheduled with relation to the raceway installation, provide the most economical means of providing the concrete envelope on a unit price per cubic yard basis, which eliminates to a great extent the highly variable labor cost factor involved in the job site mix method.

While worker jurisdictional disputes are not normally a factor in handling the concrete pour, if there is any doubt as to the situation in the local area of the installation, the supervisory personnel should ascertain the position of the crafts concerned and make any necessary arrangements in advance of commencing the work in order to avoid possible lost time due to jurisdictional disputes.

Normally, the reinforcing of concrete envelope is not required;

however, some contract specifications require reinforcing in the case of underground raceway systems extending under railroad tracks and highways. In such cases advance arrangements must be made for its provision.

Some specifications allow or require the use of a conduit "pusher" for installing single runs of conduit under railroad tracks or narrow paved roads. This eliminates the necessity for reinforced concrete.

CONDUCTORS IN UNDERGROUND RACEWAYS

Conductors installed in underground raceway systems fall into two general categories:

1. Groups of individual conductors
2. Multiconductor cables

The category used will depend on the following:

1. The contract specification requirements
2. The relative material and labor installation cost
3. The adaptability to the required use or equipment to be connected
4. The availability

A difference in cost may be expected between high- and low-voltage cables of the same wire size.

In the past, copper wire has been used almost exclusively for underground wiring systems because of copper's low resistance. Aluminum conductors have the advantage of lightness, but since aluminum has a higher resistance than copper, aluminum conductors would be larger than copper cable, requiring larger ducts or conduit. In direct burial installations, aluminum wire is now used extensively, but in conduit or duct systems, copper is still used mostly, because the ducts are usually more costly than the cable and so must carry as much power as possible.

The installation of potheads on the ends of lead-sheathed and other high-voltage cables serves to exclude moisture from the ends of the cable, protect the cable against mechanical injury, and provide terminals for the cable conductors. When dealing with potheads and the splicing of high-voltage cable, only qualified cable splicers should be used to perform the work.

Large cables are ordered in lengths to fit the duct or trench sections, plus an allowance for slack and splicing. Smaller cables are shipped on standard reels and are cut on the job after they have been installed in the ducts or laid in the trench. Cables may be pulled by hand, although the use of a cable-pulling machine or winch is the most desirable method. Sometimes a saving in labor can be realized by ordering the cables of three- or four-wire systems made up on one reel, so that all cables are pulled together.

To prevent the spread of damaging arcs from a cable fault to adjacent cables in conduits, manholes, and handholes, the cables are wrapped with an asbestos tape saturated with sodium silicate, or each cable may have a cement covering applied in the manhole.

18

Grounding of Systems

Each and every electrical system is required to be grounded in a manner prescribed by the NE code and, in some cases, local ordinances.

All pole-line operating equipment must be grounded as a safety factor. In some installations it is also required that the conductors of a de-energized line be grounded also. When gang-operated pole top disconnecting switches are used, a double-throw type of switch unit will automatically ground the circuit conductors when the circuit blades are in an open position.

The neutral of the transformer secondary circuit must also be grounded. The ground contact with the earth is accomplished either with the use of a pole-type grounding plate placed on the bottom of the pole before it is set, by means of a driven ground rod, or, in some instances, by stapling a length of coiled bare copper wire on the bottom of the pole. A bare or weatherproof insulated copper grounding conductor is installed on the pole from the ground plate or rod to the topmost level of equipment to be grounded. This conductor is usually protected for its entire exposed distance on the pole with a small size of wood pole molding.

In general, the electrical wiring system at the point of usage is grounded at one point, at the service entrance equipment location. The grounding is accomplished by using a single copper conductor, which is connected at one end to a cold-water pipe or a driven ground rod. The other end of the grounding conductor is connected to the service entrance metal enclosure by an appropriate grounding lug.

When the electric service includes a neutral conductor, it is grounded at the service entrance equipment location, usually through a solid neutral block, by means of a jumper conductor connected between the neutral block and the service raceway grounding conductor.

The feeder and branch circuit metallic raceways and the metal enclosures of panelboards and distribution centers are considered to be adequately grounded when they are mechanically connected to each other and to the service entrance equipment metal enclosure.

Where conduit enters a transformer or switchgear which is installed between the incoming conduit and the point where the ground connection is made, it is unwise to rely entirely on locknuts and bushings for electrical continuity. Electrical disturbances may occur in the wiring system that call for a good ground connection at this point. To ensure continuity of the ground, a bonding jumper is usually installed between the outside conduit and the ground connection. Bonding bushings are normally used for this purpose. Such use is especially important in cabinets having only concentric knockouts. Threaded hubs are better.

All wiring systems should be properly grounded, and the resistance to ground should be as low as possible. To ensure a good ground, clean the grounding electrode (ground rod, ground plate, water pipe, etc.), removing dirt, paint, rust, or other substances which introduce resistance to the flow of electricity. Tighten all ground clamps to the conduit and grounding electrode firmly and carefully. When using underground wiring systems, all equipment placed in manholes—as well as the metallic sheaths of conductors—should be grounded. On most projects, it is usually required to provide one or more ground rods in manholes for the attachment of grounding conductors irrespective of whether or not the equipment is initially placed in the manhole. Work crews must ascertain the grounding requirements for each individual project and provide for the installation of the necessary ground rods, ground clamps, grounding conductors, wire connectors, fastening straps, and the like.

GROUNDING OF TRANSFORMERS

In most cases, the core of the transformer as well as the coil assembly and case should be permanently and adequately grounded.

Grounding is necessary to remove static electricity and also as a precautionary measure in case the transformer windings accidentally come in contact with the core or enclosure. Installers should make certain that the flexible grounding jumper between the core and coil assembly and case is intact or that the core and coil assembly is directly grounded from the core clamp through a flexible lead. Of course, as mentioned previously, be certain that grounding and bonding meet the NE Code requirements and also local codes, where applicable.

The tank of every power transformer should be grounded to eliminate the possibility of obtaining static shocks from it or being injured by accidental grounding of winding to case. A grounding lug is provided on the base of most transformers for the purpose of grounding the case and fittings.

The NE Code specifically states the requirements of grounding and should be followed in every respect. Furthermore, there are certain advisory rules that are recommended by manufacturers that provide additional protection beyond that of the code. In general, the code requires that separately derived alternating current systems be grounded, as stated in Article 250-26:

Grounding Separately Derived Alternating Current Systems. A separately derived ac system that is required to be grounded by Section 250-5 shall be grounded as specified in (a) through (d) below.

(a) Bonding Jumper. A bonding jumper, sized in accordance with Section 250-79(c) for the derived phase conductors, shall be used to connect the equipment grounding conductors of the derived system to the grounded conductor. Except as permitted by Exception No. 4 or Section 250-23(a), this connection shall be made at any point on the separately derived system from the source to the first system disconnecting means or overcurrent device; or it shall be made at the source of a separately derived system which has no disconnecting means or overcurrent devices.

Exception: The size of the bonding jumper for a system that supplies a Class 1 remote control or signaling circuit, and is derived from a transformer rated not more than 1000 volt-amperes, shall not be smaller than the derived phase conductors and shall not be smaller than No. 14 copper or No. 12 aluminum wire.

(b) Grounding Electrode Conductor. A grounding electrode conductor, sized in accordance with Section 250-94 for the derived phase conductors, shall be used to connect the grounded conductor of the derived system to the grounding electrode as specified in (c) below. Except as permitted by Exception of No. 4 of Section 250-23(a), this connection shall be made at any point on the separately derived system from the source to the first system disconnecting means or overcurrent device; or it shall be made at the source of a separately derived system which has no disconnecting means or overcurrent devices.

Exception: A grounding electrode conductor shall not be required for a system that supplies a Class 1 remote control or signaling circuit, and is derived from a transformer rated not more than 1000 volt-amperes, provided the system grounded conductor is bonded to the transformer frame or enclosure by a jumper sized in accordance with the Exception for (a), above, and the transformer frame or enclosure is grounded by one of the means specified in Section 250-57.

(c) Grounding Electrode. The grounding electrode shall be as near as practical to and preferably in the same area as the grounding conductor connection to the system. The grounding electrode shall be: (1) the nearest available effectively grounded structural metal member of the structure; or (2) the nearest available effectively grounded metal water pipe; or (3) other electrodes as specified in Sections 250-81 and 250-83 where electrodes specified by (1) or (2) above are not available.

(d) Grounding Methods. In all other respects, grounding methods shall comply with requirements prescribed in other parts of the NE Code.

Glossary

Accessible (as applied to equipment): admitting close approach because not guarded by locked doors, elevation, or other effective means.

Accessible (as applied to wiring methods): capable of being removed or exposed without damaging the building structure or finish, or not permanently closed in by the structure or finish of the building.

Alternating current: current (ac) which reverses direction rapidly, flowing back and forth in the system with regularity. This reversal of current is due to the reversal of voltage which occurs at the same frequency. In alternating current, any one wire is first positive, then negative, then positive, and so on.

Alternator: an electric generator designed to supply alternating current. Some types have a revolving armature and other types a revolving field.

Ampacity: current-carrying capacity expressed in amperes.

Ampere: The unit of measurement for electric current. It represents the rate at which current flows through a resistance of 1Ω by a pressure of 1 V.

Amplitude: the maximum instantaneous value of an alternating voltage or current. It is measured in either the positive or negative direction.

Approved: acceptable to the authority enforcing the code.

Automatic: self-acting, operating by its own mechanism when actuated by some impersonal influence, such as a change in current strength, pressure, temperature, or mechanical configuration.

Bonding jumper: a reliable conductor used to ensure the required electrical conductivity between metal parts required to be electrically connected.

Bus bar: the heavy copper or aluminum bar used on switchboards to carry current.

Capacitor or condenser: an electrical device that causes the current to lead the voltage, opposite in effect to inductive reactance. They are used to neutralize the objectional effect of lagging (inductive reactance) which overloads the power source. Also acts as a low-resistance path to ground for currents of radio frequency, thus effectively reducing radio disturbance.

Circuit breaker: a device designed to open and close a circuit by nonautomatic means and to open the circuit automatically on a predetermined overload of current, without injury to itself when properly applied within its rating.

Circular mil: the area of a circle $\frac{1}{1000}$ in. in diameter. The area of electrical conductors is usually measured in circular mils; that is, 500,000 circular mils (500 MCM), etc.

Commutator: device used on electric motors or generators to maintain unidirectional current.

Conductor: substances that offer little resistance to the flow of electric current. Silver, copper, and aluminum are good conductors, although no material is a perfect conductor.

Connector, pressure (solderless): a connector that establishes the connection between two or more conductors or between one or more conductors and a terminal by means of mechanical pressure and without the use of solder.

Continuous load: a load in which the maximum current is expected to continue for 3 hr or more.

Controller: a device, or group of devices, that serves to govern in some predetermined manner the electric power delivered to the apparatus to which it is connected.

Current: the flow of electricity in a circuit. It is expressed in amperes and represents an amount of electricity.

Cycle: one complete period of flow of alternating current in both directions. One cycle represents 360°.

Demand factor: in any system or part of a system, the ratio of the maximum demand of the system, or part of the system, to the total connected load of the system, or part of the system under consideration.

Disconnecting means: a device, a group of devices, or other means whereby the conductors of a circuit can be disconnected from their source of supply.

Duty, intermittent: a requirement of service that demands operation for alternate intervals of (1) load and no load, (2) load and rest, or (3) load, no load, and rest.

Duty, periodic: a type of intermittent duty in which the load conditions regularly recur.

Duty, short time: a requirement of service that demands operations at loads and for intervals of time, both of which may be subject to wide variation.

Efficiency: the name given to ratio of output to input.

Electrical generator: a machine so constructed that when its rotor is driven by an engine or other prime mover, a voltage is generated.

Electrode: a conducting element used to emit, collect, or control electrons and ions.

Feedback: the process of transferring energy from the output circuit of a device back to its input.

Feeder: the conductors between the service equipment, or the generator switchboard of an isolated plant, and the branch circuit overcurrent device.

Frequency: frequency of alternating current is the number of cycles per second. A 60-Hz alternating current makes 60 complete cycles of flow back and forth (120 alternations) per second. A conventional alternator has an even number of field poles arranged in alternate north and south polarities. Current flows in one direction in an ac armature conductor while the conductor is passing a north pole and in the other direction while passing a south pole. The conductor passes two poles during each cycle. A frequency of 60 Hz requires the conductor to pass 120 poles/sec. In a six-pole alternator, the equivalent speed would be 20 rev/sec or 1200 rpm. In a four-pole alternator, the equivalent speed would be 30 rev/sec or 1800 rpm.

Fuse: a protective device inserted in series with a circuit.

Grounded conductor: a system or circuit conductor that is intentionally grounded.

Grounding conductor: a conductor used to connect equipment or the grounded circuit of a wiring system to a grounding electrode.

Hertz: a unit of frequency, 1 cycle/sec. Written as 50-Hz or 60-Hz current, etc.

Horsepower: the unit of power about equal to the power of a draft horse to do work for a short interval. Numerically, hp is 33,000 ft-lb/per min, that is, the ability to lift 33,000 lb to a height of 1 ft in 1 min.

Impedance: effects placed on alternating current by inductive capacitance (current lags voltage), capacitive reactance (current leads voltage), and resistance (opposes current but does not lag or lead voltage), or any combination of two. It is measured in ohms as is resistance.

Inductance: the property of a circuit or two neighboring circuits which determines how much voltage will be induced in one circuit by a change of current in either circuit.

Insulator: substances that offer great resistance to the flow of electric current such as glass, porcelain, paper, cotton, enamel, and paraffin are called insulators because they are practically nonconducting. However, no material is a perfect insulator.

kVA the abbreviation of kilovolt-amperes, which is the product of the volts times the amperes divided by 1000. This term is used in rating alternating-current machinery because with alternating currents, the product of the volts times the amperes usually does not give the true average power.

kvar: the abbreviation of kilovolt-ampere reactance, which is a measurement of reactive power that generates power within induction equipment (motors, transformers, holding coils, lighting ballasts, etc.).

kW: the abbreviation for kilowatt, which is a unit of measurement of electrical power. A kilowatt (kW) equals 1000 W and is the product of the volts times the amperes divided by 1000 when used in rating dc machinery. Also the term used to indicate true power in an ac circuit.

Kilowatt-hour: the amount of electrical power represented by 1000 W for a period of 1 hr. Thus a generator which delivered 1000 W for a period of 1 hr would have delivered 1 kW of electricity.

National Electrical Code: the National Electrical Code is sponsored by the National Fire Protection Association and is the "Bible" of all electrical workers for building construction. It is often referred to as the *NEC* or the *Code.*

Ohm: the unit of measurement of electrical resistance; it represents the amount of resistance that permits current flow at the rate of 1 A under a pressure of 1 V. The resistance (in ohms) equals the pressure (in volts) divided by the current (in amperes).

Panelboard: a single panel or group of panel units designed for assembly in the form of a single panel; includes buses and may come with or without switches and/or automatic overcurrent protective devices for the control of light, heat, or power circuits of small individual as well as aggregate capacity. It is designed to be placed in a cabinet or cutout box placed in or against a wall or partition and accessible only from the front.

Power: the rate of doing work or expending energy.

Power factor: when the current waves in an ac circuit coincide exactly in time with the voltage waves, the product of volts times amperes gives volt-amperes, which is true power in watts (or in kilowatts if divided by 1000). When the current waves lag behind the voltage, due to inductive reactance (or lead due to capacitive reactance), they do not reach their respective values at the same time. Under such conditions, the product of volts and amperes does not give true average watts. Such a product is called volt-amperes or apparent watts. The factor by which apparent watts must be multiplied to give the true watts is known as the power factor (PF). Power factor depends on the amount of lag or lead and is the percentage of apparent watts which represents true watts. With a power factor of 80%, a fully loaded 5-kVA alternator will produce 4 kW. When the rating of a power unit is stated in kVA at 80% PF, it means that with an 80%-PF load, the generator will generate its rated voltage providing the load does not exceed the kVA rating. In an engine-driven alternator, for example, with automatic voltage regulation, the kVA usually is determined by the maximum current which can flow through the windings without injurious overheating or by the ability of the engine or other prime mover to maintain the normal operating speed. A resistance load such as electric lamp bulbs, irons, toasters, and similar devices is a unity power-factor load. Motors, transformers, and various other devices cause a current wave lag which is expressed in the power factor of the load.

Raceway: any channel designed expressly for holding wire, cables, or bus bars and used solely for this purpose.

Raintight: so constructed or protected that exposure to a beating rain will not result in the entrance of water.

Reactance: reactance is opposition to the change of current flow in an ac circuit. The rapid reversing of alternating current tends to induce voltages that oppose the flow of current in such a manner that the current waves do not coincide in time with the voltage waves. The opposition of self-inductance to the flow of current is called inductive reactance and causes the current to lag behind the voltage which produces it. The opposition of a condenser or of capacitance to the change of ac voltage causes the current wave to lead the voltage wave. This is called capacitive reactance. The unit of measurement for either inductive reactance or capacitive reactance is the ohm.

Rectifiers: devices used to change alternating current to unidirectional current.

Relay: an electromechanical switching device that can be used as a remote control.

Remote-control circuit: any electrical circuit that controls any other circuit through a relay or an equivalent device.

Resistance: electrical resistance is opposition to the flow of electric current and may be compared to the resistance of a pipe to the flow of water. All substances have some resistance, but the amount varies with different substances and with the same substances under different conditions.

Resistor: a resistor is a poor conductor used in a circuit to create resistance which limits the amount of current flow. It may be compared to a valve in a water system.

Resonance: in a circuit containing both inductance and capacitance, a condition in which the inductive reactance is equal to and cancels out the capacitance resistance.

Service: the conductors and equipment used for delivering energy from the electricity supply system to the wiring system of the premises served.

Service drop: the overhead service conductors from the last pole, or other aerial support, to and including the splices, if any, that connect to the service entrance conductors at the building or other structure.

Service equipment: the necessary equipment, usually consisting of a circuit breaker, or switch and fuses, and their accessories, located near the point of entrance of supply conductors to a building and intended to constitute the main control and means of cutoff for the supply to that building.

Single phase: a single-phase, ac system has a single voltage in which voltage reversals occur at the same time and are of the same alternating polarity throughout the system.

Switchboard: a large single panel, frame, or assembly of panels, having switches, overcurrent, and other protective devices, buses, and usually instruments, mounted on the face or back or both. Switchboards are generally accessible from the rear as well as from the front and are not intended to be installed in cabinets.

Synchronous: simultaneous in action and in time (in phase).

Thermal cutout: an overcurrent protective device containing a heater element in addition to and affecting a renewable fusible member which opens the circuit. It is not designed to interrupt short-circuit currents.

Thermal protector (as applied to motors): a protective device that is assembled as an integral part of a motor or motor compressor and that, when properly applied, protects the motor against dangerous overheating due to overload and failure to start.

Thermally protected (as applied to motors): refers to the words *thermally protected* appearing on the nameplate of a motor or motor-compressor and means that the motor is provided with a thermal protector.

Three phase: a three phase, ac system has three individual circuits or phases. Each phase is timed, and the current alternations of the first phase are $\frac{1}{3}$ cycle (120°) ahead of the second and $\frac{2}{3}$ cycle (240°) ahead of the third.

Transformer: a device used to transfer energy from one circuit to another. It is composed of two or more coils linked by magnetic lines of force.

Utilization equipment: equipment that utilizes electric energy for mechanical, chemical, heating, lighting, or other similarly useful purposes.

Ventilated: provided with a means to permit circulation of air sufficient to remove an excess of heat fumes or vapors.

Volt: the practical unit of voltage or electromotive force. One volt sends a current of 1 A through a resistance of 1Ω.

Voltage: the force, pressure, or electromotive force (emf) which causes electric current to flow in an electric circuit. Its unit of measure is the volt, which represents the amount of electrical pressure that causes current to flow at the rate of 1 A through a resistance of 1Ω. Voltage in an electric circuit may be considered as being similar to water pressure in a pipe or water system.

Voltage drop: in an electrical circuit, the difference between the voltage at the power source and the voltage at the point at which electricity is to be used. The voltage drop or loss is created by the resistance of the connecting conductors.

Voltage to ground: in grounded circuits, the voltage between the given conductor and that point or conductor of the circuit which is grounded; in ungrounded circuits, the greatest voltage between the given conductor and any other conductor of the circuit.

Watt: the unit of measurement of electrical power or rate of work; 746 W is equivalent to 1 hp. The watt represents the rate at which power is expended when a pressure on 1 V causes current to flow at the rate of 1 A. In a dc circuit or in an ac circuit at unity (100%) power factor, the number of watts equals the pressure (in volts) multiplied by the current (in amperes).

Index

244